Springer Protocols Handbooks

For further volumes:
http://www.springer.com/series/8623

Springer Protocols Handbooks collects a diverse range of step-by-step laboratory methods and protocols from across the life and biomedical sciences. Each protocol is provided in the Springer Protocol format: readily-reproducible in a step-by-step fashion. Each protocol opens with an introductory overview, a list of the materials and reagents needed to complete the experiment, and is followed by a detailed procedure supported by a helpful notes section offering tips and tricks of the trade as well as troubleshooting advice. With a focus on large comprehensive protocol collections and an international authorship, Springer Protocols Handbooks are a valuable addition to the laboratory.

Biophysical Characterization of Functional Peptides

Edited by

Vibin Ramakrishnan

Biosciences & Bioengineering, Indian Institute of Technology Guwahati, Guwahati, Assam, India

Humana Press

Editor
Vibin Ramakrishnan
Biosciences & Bioengineering
Indian Institute of Technology Guwahati
Guwahati, Assam, India

ISSN 1949-2448 ISSN 1949-2456 (electronic)
Springer Protocols Handbooks
ISBN 978-1-0716-3404-2 ISBN 978-1-0716-3405-9 (eBook)
https://doi.org/10.1007/978-1-0716-3405-9

This Humana imprint is published by the registered company Springer Science+Business Media, LLC, part of Springer Nature.
The registered company address is: 1 New York Plaza, New York, NY 10004, U.S.A.

Dedicated to my father A. S. Ramakrishnan

Preface

The goal of this book is to support the learning objectives of undergraduate and graduate students working in the broad area of peptide chemistry and biophysics.

The history of peptide chemistry started with the synthesis of glycylglycine in 1901, by Emil Fischer and Ernest Fourneau. Peptides are involved in wide variety of functions such as antimicrobials, ligands, hormones, signaling molecules, and growth factors. The discovery of solid phase peptide synthesis (SPPS) by R. B. Merrifield in 1963 revolutionized this field of science and was looked upon as a pertinent solution for many diseases. Peptides are now used in a wide variety of applications. This book essentially deals with the experimental characterizations that can verify the efficacy of using peptides as a promising biomaterial. Protocols for the biophysical characterization for the five actively pursued research areas such as drug delivery vehicles, peptide nano-assembly, bionanocatalysis, peptide/protein aggregation diseases, and peptide based antibacterials are focused. Each protocol has materials, methods, reagents, step-by-step procedures, and readily reproducible laboratory protocols.

The market for peptide therapeutics (globally) was valued at USD 25.35 billion in 2018. This figure is projected to reach USD 50.60 billion by 2026. De novo designed peptide drugs are used as replacement therapies supplementing peptide hormones. Till this time, approval for 60 peptides has been granted. Mostly, the approved drugs are used as drug target moiety for cancer, metabolic disorders, and cardiovascular diseases in countries like USA, Japan and in European Union. More trained professionals may be necessary to meet the future challenges in the design and industrial production of therapeutic peptides. This book is expected to support initiatives to build trained manpower in peptide research.

Guwahati, Assam, India — *Vibin Ramakrishnan*

Acknowledgements

Writing or editing a book was always a passion for me, but it turned out to be harder than I expected. Editing the contributions of different writers who are experts in a field has different set of challenges than writing it on my own. However, various people helped me to finish this task and I thankfully remember all of them.

I started learning about peptides and proteins systematically when I joined for my PhD at IIT Bombay. I am thankful to Professor Susheel Durani for introducing me to this exciting scientific discipline. My colleagues in Durani Laboratory, Dr. Dhanasekaran, Dr. Mohan Raja, Dr. Ranjit Ranbhor, Dr. Anil Kumar, and Dr. Kirti Patel supported me at different times in my journey with peptides.

I continued working on peptides and proteins during my postdoctoral research under the guidance of Prof. Chris Bystroff at Rensselaer Polytechnic Institute, Troy, New York. Chris influenced me in many ways, especially the manner in which he talks to students was something I adopted when I became a faculty member. Chris is a wonderful human being, I have not seen many people of his quality in my life. I also remember Yao-ming Huang of Bystroff lab. I learned from him how to optimally use resources available as open source tools and techniques for our research.

After joining IIT Guwahati, Professor Nitin Chaudhary helped me in learning experimental techniques in peptide chemistry. At MID, we are using the synthesis protocol designed by him for many years.

I am the principal investigator of Molecular Informatics and Design (MID) Laboratory at IIT Guwahati. The contributors of this book are mostly my own students. I am blessed to have really good students joining under my supervision for PhD and M.Tech. All the experiments and protocols presented in this book were part of their research work at different times.

I am thankful to my family, starting from my mother Sailaja, who has worked really hard in my early years to develop me as a student of science, to my wife Archana for sparing me from time consuming duties, normally associated with family life.

I am always accompanied by people with "beautiful minds" in my life, and I am thankful to the Master of the Universe for this blessing.

Vibin Ramakrishnan
IIT, Guwahati

Contents

Editor and Contributors

About the Editor

VIBIN RAMAKRISHNAN is a Professor of Biosciences and Bioengineering at IIT Guwahati. He did his Master's in Applied Chemistry from Cochin University of Science and Technology and Ph.D. from School of Biosciences and Bioengineering, IIT Bombay. He worked as a Postdoctoral Research Associate at Rensselaer Polytechnic Institute, New York and a Faculty Scientist at IBAB Bangalore, before joining Department of Biosciences and Bioengineering at IITG. He has published more than 50 research articles in reputed international journals and filed 10 patents, including an international patent to his credit. He is also the principal scientific advisor of two biotech companies.

The Department of Biotechnology, Govt. of India has awarded him the Innovative Young Biotechnologist Award (IYBA) in the year 2007–2008 to recognize his contributions and for innovative approach in biotechnology research. He is an editorial board member of Scientific Reports. He has more than 15 years of teaching experience in biophysics, computational biology, biotechniques, analytical biotechnology, and proteomics. He is a member of many international scientific societies such as American Chemical Society, Biophysical Society, and Protein Society.

Contributors

LAXMI RAMAN ADIL • *Biosciences & Bioengineering, Indian Institute of Technology Guwahati, Guwahati, Assam, India*

KHUSHBOO BHAGAT • *Biosciences & Bioengineering, Indian Institute of Technology Guwahati, Guwahati, Assam, India*

AMIT BHANDARI • *Biosciences & Bioengineering, Indian Institute of Technology Guwahati, Guwahati, Assam, India*

ILACKKEYA BHAVANANTHI • *Biosciences & Bioengineering, Indian Institute of Technology Guwahati, Guwahati, Assam, India*

YVONNE CHRISTIAN • *Biosciences & Bioengineering, Indian Institute of Technology Guwahati, Guwahati, Assam, India*

DEBIKA DATTA • *Department of Nanoengineering, University of California San Diego, La Jolla, CA, USA*

RUCHIKA GOYAL • *Biosciences & Bioengineering, Indian Institute of Technology Guwahati, Guwahati, Assam, India*

DIKSHITA HAZARIKA • *Biosciences & Bioengineering, Indian Institute of Technology Guwahati, Guwahati, Assam, India*

NAVEEN KUMAR • *Department of Biosciences & Bioengineering, Indian Institute of Technology, Guwahati, Assam, India*

KALPANA KUMARI • *Biosciences & Bioengineering, Indian Institute of Technology Guwahati, Guwahati, Assam, India*

GAURAV PANDEY • *Department of Molecular Medicine, Scripps Research Institute, La Jolla, CA, USA*

VIVEK PRAKASH • *Biosciences & Bioengineering, Indian Institute of Technology Guwahati, Guwahati, Assam, India*

VIBIN RAMAKRISHNAN • *Biosciences & Bioengineering, Indian Institute of Technology Guwahati, Guwahati, Assam, India*

AMAY SANJAY REDKAR • *Biosciences & Bioengineering, Indian Institute of Technology Guwahati, Guwahati, Assam, India*

JAHNU SAIKIA • *Biosciences & Bioengineering, Indian Institute of Technology Guwahati, Guwahati, Assam, India*

MOULI SARKAR • *Department of Biosciences & Bioengineering, Indian Institute of Technology, Guwahati, Assam, India*

Chapter 1

De Novo Peptide Design Tools

Mouli Sarkar and Naveen Kumar

Abstract

This chapter illustrates the utility of a toolkit consisting of five programs to design peptides and small proteins. The toolkit comprises five different programs that can be used independently or together as a pipeline for the design and computational characterization of peptides. The toolkit includes programs for checking the chirality of amino acids (ProChiral), generating contact maps (CoMa), analyzing data redundancy (DaRe), calculating the hydrogen bond potential energy (HyPE), and electrostatic interaction energy (EsInE). The programs can be downloaded from http://www.iitg.ac.in/bpetoolkit/.

Key words Peptide design, Electrostatic interaction energy, Python, Potential energy, Chirality, Contact maps, Hydrogen bonds

1 Introduction

Peptides are crucial for numerous metabolic processes and play important roles in various physiological functions. It is a short stretch of 2–50 amino acids synthesized as a result of a condensation reaction, bonded through a covalent link [1].

The name "peptide" is assigned based on the length of polypeptide chain, which in turn decided by the number of amino acid residues in the polymeric chain. As peptide chains integrate the fundamental structure of amino acids, an oligopeptide may form when there are between 10 and 20 amino acids in the chain. To synthesize a peptide chain in vivo, each amino acid has to be bonded to the amino terminus of the previous amino acid. A peptide chain is generally considered as a polypeptide when it has more than 20 amino acids and is unbranched [1]. Polypeptide chains can in principle have amino acids of both L and D chirality, but naturally occurring proteins have only L-chiral amino acids [2].

The last decade has seen significant advancements in the development of peptide drugs owing to advances in technology for design, modification, and analysis. Chemical, biological, and innovative design and delivery techniques have been used to

Vibin Ramakrishnan (ed.), *Biophysical Characterization of Functional Peptides*, Springer Protocols Handbooks,
https://doi.org/10.1007/978-1-0716-3405-9_1,

manufacture and modify peptides [3, 4]. Due to their distinctive biochemical properties and therapeutic potential as peptide-mimicking hormones, drug delivery systems, therapeutic peptides, GLP analogs, peptide-based antibiotics, and peptide-based hydrogels, have recently emerged as a distinct class of therapeutic agents [4, 5]. Therapeutic peptides have a broad range of applications, including the treatment of cancer, diabetes, gastric and intestinal diseases, and as antiviral agents [4, 6].

The methods for designing de novo peptides and exploring their characteristics are discussed in this chapter. It briefly outlines many approaches that can be used to identify several parameters required for peptide design.

2 Materials

- Linux Operating system.
- Python 2.7.
- Swiss-PDB Viewer.
- MinGW compiler.

3 Methods

3.1 Check Chirality

The chirality of a molecule may be characterized by the absence of a superimposable mirror image. All amino acids except glycine have chiral centers, which are C-alpha (CA) atoms. Checking the orientation of the side chain from the peptide backbone or the direction of the CA atom in the structure can be used to define the chirality of the amino acids. This can be achieved by analyzing the improper dihedral angle produced by the normals from the C-CA-CB and N-CA-C planes. Levorotatory amino acids have an angle with a positive value, whereas dextrorotatory amino acids have an angle with a negative value. The orientation of the side chain of the amino acid residues can be determined by using the chirality program ProChiral, as either Dextrorotatory or Levorotatory. 1GRM.pdb was used to test ProChiral, which has a combination of D and L amino acids [7, 8].

1. All the programs can be freely downloaded from the website https://www.iitg.ac.in/bpetoolkit/. The following file is a part of the package: chiral.sh, ca_Chirality.c, coord_extract.sh, ReadMe.
2. One PDB file should be placed as input to the program in that directory.
3. The following command is entered in the terminal to execute the script (see **Note 2**):

```
DVA   6     -122.479309      D
VAL   7    122.440262   L
DVA   8     -122.460045      D
TRP   9    121.519257   L
DLE   10    -123.225464      D
TRP   11    121.563477   L
DLE   12    -123.312088      D
TRP   13    121.436432   L
DLE   14    -123.453873      D
TRP   15    121.484276   L
FVA   1     122.419296   L
ALA   3     121.655899   L
DLE   4     -123.278114      D
ALA   5     121.701706   L
```

Fig. 1 Results of chirality determination of a peptide by Prochiral. The output file is "Chirality Result.txt"

```
$ bash chiral. sh
```

4. The output is a text file with the name "Chirality Result.txt," and the results are displayed in the terminal. The results are shown in Fig. 1.

3.2 CoMa: Contact Map Program

The spatial configurations of a protein can be represented in a reduced form as Contact maps. Contact maps are two-dimensional graphic representations of a three-dimensional peptide or protein structure. The two most popular types of maps are based on either the CA-CA distance or the CB-CB (CA in the case of glycine) distance, and the reported cut-off ranges from 6 to 12 Å. Parallel beta sheets, alpha helices, and anti-parallel beta sheets can be identified by analyzing the contact map. One can generalize this utility to the discovery and/or modeling of different protein folds [7].

1. The BPE Toolkit includes the CoMa program. The program can be downloaded from the website https://www.iitg.ac.in/bpetoolkit/. Files such as Extract.py, CoMa.sh, CoMa.c, map.c, and ReadMe were included in this package.
2. The following command is entered on the terminal to execute the script (see **Note 2**):

```
$ bash CoMa.sh
```

3. It will produce two output files: "Output. XMP," which may be viewed using any picture viewer, and "Output.txt," which will reveal information about the contact map of each residue.

3.3 HyPE: Hydrogen Bond Potential Energy Program

Hydrogen bonding is the most significant nonbonding interaction in biomolecules. The purpose of HyPE is to estimate the possible strength of hydrogen bonds that may form between the C=O and

```
Length between nitrogen of residue A and oxygen of residue no A is 1.89980551636
Angle between N,H,O is 4.22337224099
Hydrogen bond energy is:419.15304783
Total hydrogen bond energy is 419.15304783 Kcal/mol
```

Fig. 2 Detailed analysis of a peptide using the HyPE package and stored in "Hydrogen Bond.txt"

N-H atoms in the primary structural region of the protein [7]. The following equation is used to determine the potential energy of hydrogen bonds:

$$E_{\mathrm{hHBb}} = D_0 \left[5.\left(\frac{R0}{R}\right)^{12} - 6.\left(\frac{R0}{R}\right)^{10} \right] (\cos\theta)^2 (\cos\phi)^2 \quad (1)$$

1. PDB file is the input for the HyPE program. It is possible to use even a text file with atom coordinates extracted from the relevant PDB file (the format has to be the same).
2. The user must add hydrogen to retrieve the hydrogen coordinates of each residue (using Swiss-PDB viewer). The PDB file was used as the input file.
3. The Do and Ro parameters of the calculation provided in the Hamilton function of the program (provided by the 12-10 model) can also be edited by the user. The application also enables the user to adjust the interatomic distance threshold between "N" and "O."
4. The following command is entered to execute the script (see **Note 1**):

```
$ python HyPE.py input_file_name.PDB
```

5. The output file "Hydrogen Bond.txt" was created (Fig. 2). This provides the following details:
   (a) Length between the nitrogen of residue A and oxygen of residue no. A.
   (b) Angle between N, H, and O.
   (c) Hydrogen bond energy.
   (d) Total hydrogen bond energy.

3.4 EsInE: Electrostatic Interaction Energy

Electrostatic interactions significantly affect the functionality and stability of macromolecules with charges in aqueous solutions. Electrostatic interactions influence the binding of metal ions and charged substrates to biomolecules [9]. Electrostatic forces play a crucial role in protein–protein interactions and protein folding [7]. Once

```
electrostatic interaction energy in kilocal/mol = 566273.509618
```

Fig. 3 Result of electrostatic interaction energy calculation by EsInE package for a given peptide structure

the steric constraints are achieved, it has been studied that electrostatic forces influence interactions between biological molecules, especially those involving drugs and receptors. Electrostatic profiling can provide insights related to protein–peptide interactions [10].

EsInE software from the bPE toolkit was used to assess the electrostatic profile of the peptides. The following steps were followed to utilize the package:

1. Open a terminal and specify the "EsInE" folder path.
2. A PDB file should be placed in the directory.
3. The command below is entered in the terminal (see **Note 1**).

```
$ python EsInE.py
```

4. The PDB file name may be as follows:

```
$ 1grm.pdb
```

5. It generates an output file named "Output.txt," which provides the electrostatic energy in kcal/mol (Fig. 3).

4 Notes

1. Older version of python is used to design the tools mentioned in this chapter. It is suggested to either use an older version of python or make some changes in the scripts for successful execution.
2. Linux operating system is necessary in order to run all the programs smoothly.

Acknowledgment

The protocol was based on our previous study [7].

References

1. Forbes J, Krishnamurthy K (2022) Biochemistry, peptide. In: StatPearls. StatPearls Publishing, Treasure Island, FL
2. Ranbhor R, Ramakrishnan V, Kumar A et al (2006) The interplay of sequence and stereochemistry in defining conformation in proteins and polypeptides. Biopolymers 83:537–545
3. Saikia J, Bhat VT, Potnuru LR et al (2022) Minimalist de novo design of an artificial enzyme. ACS Omega 7:19131–19140
4. Goyal R, Jerath G, Chandrasekharan A et al (2022) Molecular hybridization combining tumor homing and penetrating peptide domains for cellular targeting. Drug Deliv Transl Res 12:1285–1292
5. Prakash V, Christian Y, Redkar AS et al (2022) Antibacterial hydrogels of aromatic tripeptides. Soft Matter 18:6360–6371
6. Jerath G, Darvin P, Christian Y et al (2022) Delivery of small molecules by syndiotactic peptides for breast cancer therapy. Mol Pharm 19:2877–2887
7. Jerath G, Hazam PK, Ramakrishnan V (2014) bPE toolkit: toolkit for computational protein engineering. Syst Synth Biol 8:337–341
8. Berman HM, Westbrook J, Feng Z et al (2000) The protein data bank. Nucleic Acids Res 28: 235–242
9. André I, Kesvatera T, Jönsson B et al (2004) The role of electrostatic interactions in calmodulin-peptide complex formation. Biophys J 87:1929–1938
10. Wang F, Gomez-Escudero A, Ramireddy RR et al (2013) Electrostatic control of peptide side-chain reactivity using amphiphilic homopolymer-based supramolecular assemblies. J Am Chem Soc 135:14179–14188

Chapter 2

Molecular Docking

Amit Bhandari and Vibin Ramakrishnan

Abstract

The rise in bioactive peptide discovery has driven the need to screen peptides in the early phases of drug discovery and development. Molecular docking is a widely used tool to assess protein–peptide interactions. Multiple methods of executing protein–peptide docking can be broadly classified as template-based and template-independent. HPEPDOCK is a template-independent docking web server, capable of globally sampling peptide binding orientations. This chapter details the protocol for using the HPEPDOCK web server for protein–peptide docking.

Key words Molecular docking, HPEPDOCK, Protein–peptide interaction, Template-independent docking

1 Introduction

Molecular docking has emerged as an essential computational tool in the last few decades to aid the early phases of drug discovery. Docking tools help to model the interactions between protein and small molecules at the atomic level [1]. The advent of peptides as therapeutic candidates in recent times demands computational tools for studying protein–peptide interaction. Also, protein–peptide interactions are essential for various biological processes like cell signaling, cell transport, apoptotic pathways, and immune response [2]; they account for almost 40% of total protein–protein interactions in the cells [3]. Computational tools designed for predicting protein–small molecule interactions are unsuitable for protein–peptide interaction as peptides are much more flexible and bigger than small molecular ligands [4]; peptides generally bind at the surface of proteins rather than in deeply buried sites which often leads to transient interactions [2]. Peptides do not have stable structures before forming complex, and their structure is not known a priori [3]. The importance of protein–peptide interactions and their increasing use in the drug discovery and structure-based

Vibin Ramakrishnan (ed.), *Biophysical Characterization of Functional Peptides*, Springer Protocols Handbooks,
https://doi.org/10.1007/978-1-0716-3405-9_2,

drug design process [5, 6] have led to the development of various docking methods entirely dedicated to protein–peptide docking.

The algorithm for protein–peptide docking can be mainly divided into two categories: template-based docking and template-independent docking [7]. Both of these approaches have different levels of prediction accuracy and input requirement. Template-based docking method applies knowledge from already present protein–peptide complex structures in the databases to find similar complexes for a given target complex, such as GalaxyPepDock [8] and PepComposer [9]. Template-independent docking methods do not require any template and are further subdivided into local and global docking. The local docking method requires defining the binding site by the user, like HADDOCK [10], DINC [11], and AutoDock CrankPep [12], whereas global docking methods do not require any such information [4]. The global docking method searches the entire surface of the protein to predict the binding site of the peptide. Various docking programs are available for global docking, like MDockPeP [13], CABS-dock [14], pepATTRACT [15], AnchorDock [16], ClusProPeptiDock [17], HPEPDOCK [18], and PIPER-FlexpepDock [19].

Template-based docking uses already-known structures as scaffolds to build a model of the complex. These tools are based on the theory that conserved binding sites on protein and peptide motifs are responsible for stabilizing many protein–peptide complexes [20]. Therefore, homologous proteins with already defined protein–peptide interactions can be used as a template to predict protein–peptide complex structures [21]. The local docking method searches for binding poses in the proximity of the user-defined site. The accuracy of local docking methods is as good as the information provided by the user on the binding site. Global docking methods extensively search for peptide binding sites and poses on the protein surface. No prior information on peptide structure and binding site is provided in global docking tools. These tools can predict peptide confirmations based on a user-provided template using various strategies like simulating peptide folding in solution [16] and lacing peptide sequences into predefined template conformations [22]. Global docking is the most computationally intensive process out of all the methods discussed so far in this chapter.

Unlike most of the protein-small molecule ligand docking experiments, where the binding site is generally known, peptide–protein binding information is not available in most cases. Therefore, protein–peptide docking requires a global search around whole protein with a peptide conformation for putative binding sites. Many global protein-peptide docking algorithms have been developed in recent years for predicting protein–peptide complexes [14, 15, 18, 19, 21, 23–28].

In this chapter, we describe a detailed protocol for a global protein-peptide docking tool. The docking tool was selected based on the following criteria:

1. It should be freely available to the public.
2. Only protein–peptide docking tools were considered (as they give better results than protein–protein or protein–small molecule docking tools [28]).
3. Should accept a wide range of peptides for docking (≤20 amino acids).
4. Should have performed well in benchmarking studies.

The protein–peptide docking tool described in this chapter is HPEPDOCK [18]. It is based on an algorithm that efficiently considers peptide flexibility in docking. HPEPDOCK has performed better than other global docking methods like CABS-dock, MDockPep, and pepATTRACT in the benchmarking study [28]. It has also achieved the best binding mode predictions for protein–peptide complexes, with peptides having sequence lengths of up to 20 amino acids.

2 Computational Method

2.1 HPEPDOCK Docking Methodology

The web server for HPEPDOCK can be accessed through the URL: http://huanglab.phys.hust.edu.cn/hpepdock/. It is freely available to all users without registration. The server has an interface for the HPEPDOCK method and is well documented with the Help and Output example. HPEPDOCK considers peptide flexibility and generates an ensemble of peptide conformations employing the MODPEP program [29]. Using MDock [30] docking protocol, these peptide conformations are globally docked against the whole protein.

Inputs for the HPEPDOCK server are receptors and peptides. They can be both sequences and structures. Depending on the input provided, the server will prepare the molecules. If the structure of the receptor is provided, then it is used as it is. Else, if the sequence is the input, its three-dimensional structure is constructed through homology modelling. The sequence similarity search is performed using the HHSuite package [31] against the PDB sequence database to find the homologous template. The template with the highest sequence coverage, similarity, and resolution is selected. The template sequence is converted to the 3D structure using MODELLER [32]. If peptides are provided in the form of structures, they are converted to a sequence. MODPEP program then generates an ensemble of peptide conformations. The ensemble of peptides is then docked on the whole receptor protein using a modified version of the MDock [30] docking protocol. MDock protocol is modified for protein-peptide docking by two aspects. First, using a reduced peptide model during the generation of different orientations for peptides. In the reduced peptide model, an amino acid is represented by two pseudo-atoms, the CA atom and the center of mass of the rest of the atoms, to reduce memory consumption and improve processing speed. Second, the iterative knowledge-based scoring function is used in place of the original protein-ligand scoring function. After generating peptides orientation with the reduced peptide model, they are converted back to all-atom models for binding energy scoring. SIMPLEX minimization algorithm is used to optimize the binding energy scores. For docking, the DMS method is used to generate the molecular surface of the receptor proteins, and SPHGEN is used to generate the negative images of the molecular shape represented in the form of sphere points. Docking is performed by matching the sphere points of the receptor proteins with the peptide atoms for generating putative peptide binding orientations. After finishing the docking, the final docking models are provided on the web page. The top ten binding models can be viewed instantly, whereas other models can be downloaded.

2.2 Running the HPEPDOCK Web Server

HPEPDOCK is available as a web server [18]. No registration is required to access the server. Users can run docking experiments on the HPEPDOCK server by providing the following inputs:

1. Protein 3D model in PDB format, or the user can also provide the PDB ID: Chain ID of the protein or protein sequence in FASTA file format. For the input sequence, the server will generate a 3D model of protein based on the homologous template in the database. (For additional input hint, *see* **Note 1**).
2. A peptide sequence in FASTA format, PDB file of the peptide and PDB: Chain ID of the peptide. (For additional input hint, *see* **Note 2**). Figure 1 shows the user interface of the HPEPDOCK docking server.

2.2.1 Advanced Options

The HPEPDOCK web server provides various advanced options for protein–peptide docking. Each option is described in detail as follows:

1. Users can provide disulfide bond-forming residue or residue proximity for a cyclic peptide. The input number should be the count number of the amino acid in the peptide sequence.
2. The default number of peptide binding models to output is 100. Users can input the required number of binding modes.
3. The user can also provide the binding site in the protein sequence for local docking. The binding residue needs to be provided in the following manner:

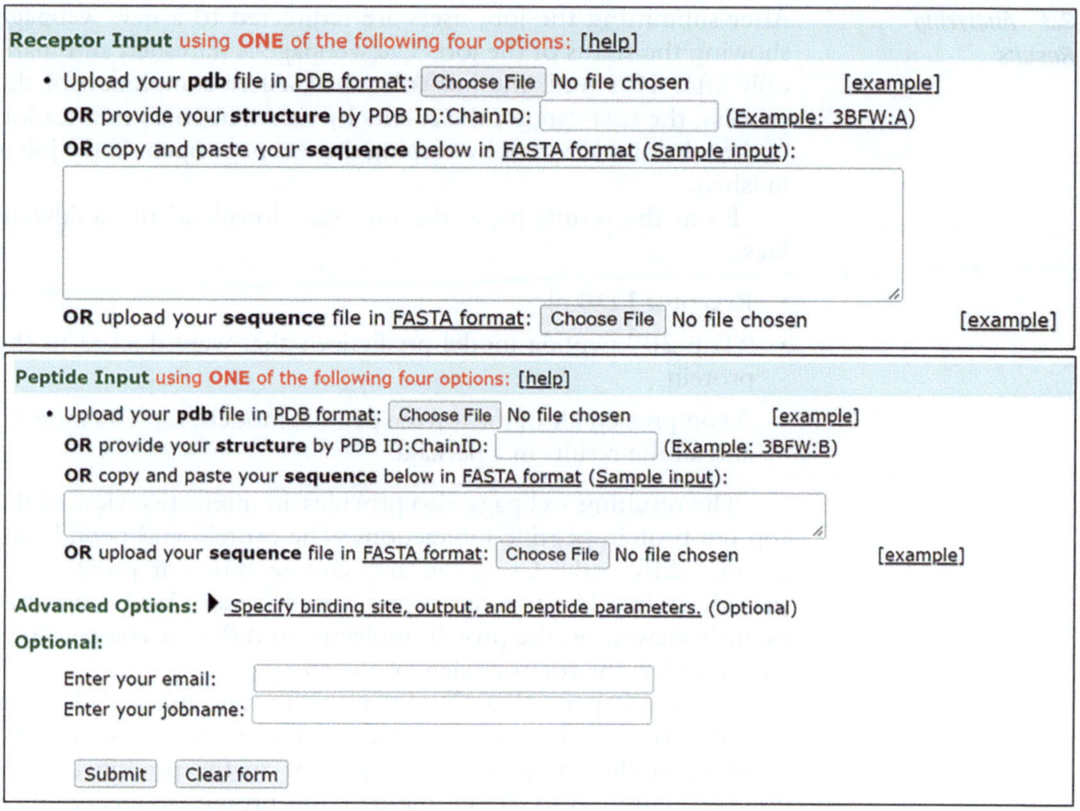

Fig. 1 Screenshot of HEPEPDOCK web server. The panel above shows the user interface for uploading protein, and the panel below shows the peptide input panel

```
"number(s): chain."

Example 195: A, 303-319:B
```

This stands for the 195th residue of chain A and residue from 303 to 319 amino acids in chain B.

The residue should be separated by a comma and should have a chain identifier, as mentioned.

4. The binding site in the receptor can also be provided in the form of a reference PDB file around which the peptide will be docked.
5. If the user wants to perform docking only with the provided peptide structure and not with different confirmations, then the option for rigid docking is also provided in this section.
6. User can provide their email ID and also job name to get a notification when the docking run is complete.

2.3 Analyzing Results

After submitting the job, users are redirected to a new webpage showing the status of the job. The webpage is refreshed automatically after 10 s until the job is finished. After completion of the process, the user can get the results by either an email notification of the job completion or by keeping the webpage open till the job is finished.

From the results page, the user can download the following files:

- Receptor PDB file.
- **"Top 20"** peptide model predictions that were docked to the protein.
- A compressed file of the top ten peptide models, top 100 models, and all the results in a package.

The resulting webpage also provides an interactive view of the top ten protein-peptide interactions. The protein and peptide are colored differently. Users can also choose different protein and peptide styles. The top ten peptide models can also be simultaneously viewed on the protein molecule in different colors. Users can also view the rotatory view of the complex.

The result page (Fig. 2) also provides the rank and docking energy score of the top ten models. These scores are not the absolute binding affinities of the peptide to the protein but are relative figures. Also, if the input is the protein sequence, then information on the template model, sequence identity, and multiple sequence alignment is also provided.

3 Methods

3.1 Case Study with Human Alpha-Thrombin with Its Inhibitor Peptide Hirudin

This case study presents an example of a protocol for HEPEP-DOCK docking performed with default settings in the web server. The receptor protein is the human alpha-thrombin, a serine protease activated by proteolysis of zymogen prothrombin (PDB: B-chain of 1A3B). It plays a major role in clot formation [33]. The peptide is Hirudin, a polypeptide secreted by the salivary gland of *Hirudo medicinalis.* It is a potent antithrombic peptide. Disulfide bonds at the N-terminal and acidic amino acids at the C-terminal help hirudin bind to thrombin and mediate its anticoagulation activity [34]. Hirudin-alpha-thrombin complex plays a role in the inhibition of the clotting process. The complex is well explored, and its structure is elucidated (PDB ID: 1A3B).

3.1.1 Job Submission with HPEPDOCK

To dock the peptide onto the protein, the job may be submitted in the following manner:

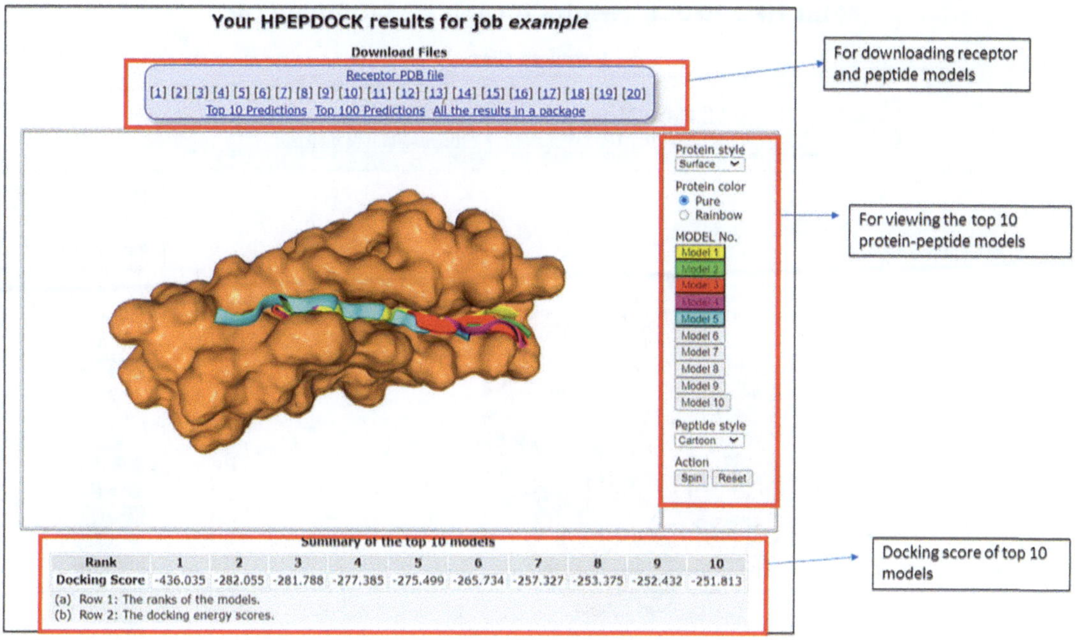

Fig. 2 Screenshot of the result page of protein–peptide after docking is complete

1. In the "Receptor Input" tab, the sequence of the protein is the input in the FASTA format:

```
>1H8D_H|Chain B [auth H] |ALPHA-THROMBIN (LARGE SUBUNIT)|Homo
sapiens (9606)

IVEGSDAEIGMSPWQVMLFRKSPQELLCGASLISDRWVLTAAHCLLYPPWDKNF-
TENDLLVRIGKHSRTRYERNIEKISMLEKIYIHPRYNWRENLDRDIALMKLKKPVAFS-
DYIHPVCLPDRETAASLLQAGYKGRVTGWGNLKETKGQPSVLQVVNLPIVERPVCKD-
STRIRITDNMFCAGYKPDEGKRGDACEGDSGGPFVMKSPFNNRWYQMGIVSWGEGCDRDG-
KYGFYTHVFRLKKWIQKVIDQF
```

2. The peptide sequence is the next input in FASTA format in the peptide tab:

```
>Hirudin Peptide sequence

GDFEEIPEEYL
```

3. Rest all the parameters were set to the default values in web server settings.
4. Basic information like job number is provided while the job is running.
5. The results were downloaded and viewed using UCSF Chimera [35]. The top ten models can be viewed on the webserver page.

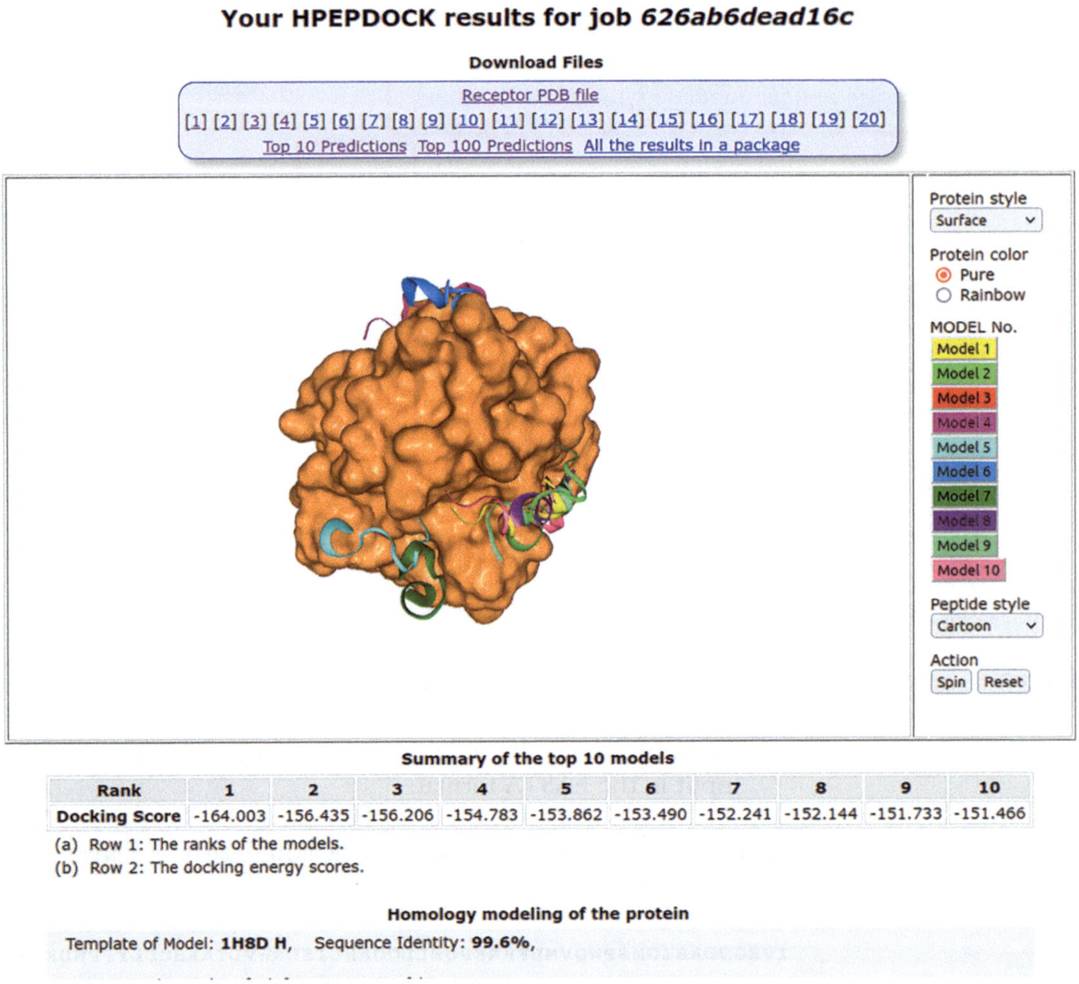

Fig. 3 A snapshot of the result page. Receptor protein is displayed in surface style and top ten peptides with different colors in ribbon style

A snapshot of the results may be seen in Fig. 3 with the assigned job number (for example, job number 626ab6dead16c).

6. The results were further analyzed on UCSF Chimera interactive visualization tools. Figure 4 shows the top-ranked, predicted binding model (blue), and native binding of the peptide to protein (red).

4 Notes

1. If no template for the provided sequence is found, then no docking will be performed. Users can also upload a file with the FASTA sequence of the protein. Modified or unnatural amino

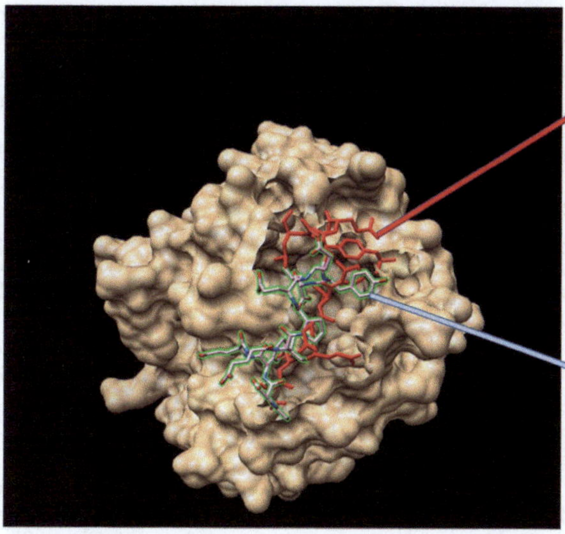

Fig. 4 Top-ranked (predicted) binding model (blue) and native binding of the peptide onto the protein (red)

acids are not allowed. The maximum length of the protein sequence is 2000 amino acids.

2. There is no limit on the length of the peptide sequence, but for better accuracy, the size should be under 30 amino acids.

References

1. Meng X-Y, Zhang H-X, Mezei M et al (2011) Molecular docking: a powerful approach for structure-based drug discovery. Curr Comput Aid Drug Des 7:146–157
2. Petsalaki E, Russell RB (2008) Peptide-mediated interactions in biological systems: new discoveries and applications. Curr Opin Biotechnol 19:344–350
3. Wang J, Alekseenko A, Kozakov D et al (2019) Improved modeling of peptide-protein binding through global docking and accelerated molecular dynamics simulations. Front Mol Biosci 6:112
4. Ciemny M, Kurcinski M, Kamel K et al (2018) Protein–peptide docking: opportunities and challenges. Drug Discov Today 23:1530–1537
5. Ozkirimli E, Sariyar B (2012) Protein-peptide interactions revolutionize drug development. IntechOpen, London. https://www.intechopen.com/chapters/39202
6. Wang X, Ni D, Liu Y, Lu S (2021) Rational design of peptide-based inhibitors disrupting protein-protein interactions. Front Chem 9:682675
7. Zhou P et al (2018) Hierarchical flexible peptide docking by conformer generation and ensemble docking of peptides. J Chem Inf Model 58:1292. https://pubs.acs.org/doi/10.1021/acs.jcim.8b00142
8. Lee H, Heo L, Lee MS et al (2015) GalaxyPepDock: a protein–peptide docking tool based on interaction similarity and energy optimization. Nucleic Acids Res 43:W431–W435
9. Obarska-Kosinska A, Iacoangeli A, Lepore R et al (2016) PepComposer: computational design of peptides binding to a given protein surface. Nucleic Acids Res 44:W522–W528
10. Spiliotopoulos D, Kastritis PL, Melquiond ASJ et al (2016) dMM-PBSA: a new HADDOCK scoring function for protein-peptide docking. Front Mol Biosci 3:36
11. Dhanik A et al (2013) DINC: a new AutoDock-based protocol for docking large ligands. BMC Struct Biol 13:S11. https://bmcstructbiol.biomedcentral.com/articles/10.1186/1472-6807-13-S1-S11
12. Zhang Y, Sanner MF (2019) AutoDock CrankPep: combining folding and docking to predict protein–peptide complexes. Bioinformatics 35:5121–5127

13. Xu X et al (2018) MDockPeP: an ab-initio protein–peptide docking server. J Comput Chem 38:2409. https://onlinelibrary.wiley.com/doi/10.1002/jcc.25555
14. Kurcinski M et al (2019) CABS-dock standalone: a toolbox for flexible protein–peptide docking. Bioinformatics 35:4170. https://academic.oup.com/bioinformatics/article/35/20/4170/5378705
15. de Vries SJ et al (2017) The pepATTRACT web server for blind, large-scale peptide-protein docking. Nucleic Acids Res 45:W361. https://pubmed.ncbi.nlm.nih.gov/28460116/
16. Ben-Shimon A, Niv MY (1993) (2015) AnchorDock: blind and flexible anchor-driven peptide docking. Struct Lond Engl 23:929–940
17. Porter KA, Xia B, Beglov D et al (2017) ClusPro PeptiDock: efficient global docking of peptide recognition motifs using FFT. Bioinformatics 33:3299–3301
18. Zhou P, Jin B, Li H et al (2018) HPEPDOCK: a web server for blind peptide–protein docking based on a hierarchical algorithm. Nucleic Acids Res 46:W443–W450
19. Alam N, Goldstein O, Xia B et al (2017) High-resolution global peptide-protein docking using fragments-based PIPER-FlexPepDock. PLoS Comput Biol 13:e1005905
20. Trabuco LG, Lise S, Petsalaki E et al (2012) PepSite: prediction of peptide-binding sites from protein surfaces. Nucleic Acids Res 40: W423–W427
21. Lee H, Seok C (2017) Template-based prediction of protein-peptide interactions by using GalaxyPepDock. In: Schueler-Furman O, London N (eds) Modeling peptide-protein interactions: methods and protocols. Springer, New York, NY, pp 37–47
22. de Vries SJ, Rey J, Schindler CEM et al (2017) The pepATTRACT web server for blind, large-scale peptide–protein docking. Nucleic Acids Res 45:W361–W364
23. Kurcinski M, Pawel Ciemny M, Oleniecki T et al (2019) CABS-dock standalone: a toolbox for flexible protein–peptide docking. Bioinformatics 35:4170–4172
24. Morris GM, Huey R, Lindstrom W et al (2009) AutoDock4 and AutoDockTools4: automated docking with selective receptor flexibility. J Comput Chem 30:2785–2791
25. Liu Z, Dominy BN, Shakhnovich EI (2004) Structural mining: self-consistent design on flexible protein–peptide docking and transferable binding affinity potential. J Am Chem Soc 126:8515–8528
26. Raveh B, London N, Zimmerman L et al (2011) Rosetta FlexPepDock ab-initio: simultaneous folding, docking and refinement of peptides onto their receptors. PLoS One 6: e18934
27. Kurcinski M, Jamroz M, Blaszczyk M et al (2015) CABS-dock web server for the flexible docking of peptides to proteins without prior knowledge of the binding site. Nucleic Acids Res 43:W419–W424
28. Weng G et al (2020) Comprehensive evaluation of fourteen docking programs on protein–peptide complexes. J Chem Theory Comput 16:3959–3969. https://pubs.acs.org/doi/10.1021/acs.jctc.9b01208
29. Yan Y, Zhang D, Huang S-Y (2017) Efficient conformational ensemble generation of protein-bound peptides. J Cheminformatics 9: 59
30. Huang S-Y, Zou X (2007) Ensemble docking of multiple protein structures: considering protein structural variations in molecular docking. Proteins 66:399–421
31. Steinegger M et al (2019) HH-suite3 for fast remote homology detection and deep protein annotation. BMC Bioinformatics 20:473. https://bmcbioinformatics.biomedcentral.com/articles/10.1186/s12859-019-3019-7
32. Eswar N, Webb B, Marti-Renom MA et al (2006) Comparative protein structure modeling using modeller. Curr Protoc Bioinformatics 5:Unit-5.6
33. Richardson JL, Kröger B, Hoeffken W et al (2000) Crystal structure of the human α-thrombin–haemadin complex: an exosite II-binding inhibitor. EMBO J 19:5650–5660
34. Junren C, Xiaofang X, Huiqiong Z et al (2021) Pharmacological activities and mechanisms of hirudin and its derivatives - a review. Front Pharmacol 12:660757
35. Pettersen EF, Goddard TD, Huang CC et al (2004) UCSF Chimera—a visualization system for exploratory research and analysis. J Comput Chem 25:1605–1612

Chapter 3

Molecular Dynamics Simulations

Amay Sanjay Redkar and Vibin Ramakrishnan

Abstract

Molecular simulations aim to supplement the design of new peptide molecules and the prediction of their conformational landscape. The response of such molecules to various physical and chemical perturbations can also be studied. This diversity in investigation options and relatively economical execution makes simulation studies one of the most important preliminary steps in peptide design and characterization. Groningen Machine for Chemical Simulations (GROMACS) is a versatile package for performing molecular dynamics simulations and has been extensively applied for a wide variety of simulations. This protocol will support the efforts of performing molecular dynamics simulations to study peptides using the GROMACS package.

Key words Molecular dynamics simulations, Peptide conformational analysis, Peptide design and characterization, GROMACS

1 Introduction

Molecular dynamics simulations are an attempt to map the conformational microstates of molecules in solution by employing a very well-established force field. Molecular motion and its time-dependent evolution are determined by integrating Newton's laws of motion across microstates. The forces are the negative derivatives (slopes) of the potential energy function, which depend on the position of atoms in a given molecule [1]. Molecular dynamics simulations provide detailed information on the conformational changes of peptides, proteins, and nucleic acids. These methods are now routinely used to investigate biological molecules and supplement the efforts to determine structure from X-ray crystallographic and spectroscopic experiments like nuclear magnetic resonance (NMR).

Peptides are typically flexible molecules that can attain a diverse array of conformations with equally diverse effects in their function. Unlike typical drug discovery methods, rigid structures are limited in application due to these conformational fluctuations. Hence,

Vibin Ramakrishnan (ed.), *Biophysical Characterization of Functional Peptides*, Springer Protocols Handbooks,
https://doi.org/10.1007/978-1-0716-3405-9_3,

molecular dynamics simulations are appropriate for deducing the conformations and validating their stability alongside the functions of a peptide in silico.

Peptides' molecular dynamics have been employed to identify unique folds in peptide conformations [2]. Molecular dynamics experiments have also been employed to uncover associations between amino acid stereochemistry encompassing the peptide and protein chain, sequence, and its conformations [3–5]. Antimicrobial peptides have also been developed using peptide membrane simulations, incorporating unnatural amino acids and varied tacticity [6–9]. Peptide membrane simulations have also been deployed to design peptide-based drug delivery systems specifically targeting tumor cells using cell-penetrating peptides (CPPs) [10–13].

GROMACS is a molecular dynamics simulation package, typically run on the Linux operating system to simulate macromolecules such as proteins, nucleic acids, and lipids. It is currently one of the fastest molecular dynamics simulation toolkits, which is powerful on its own but also integrates with other third-party software. It is also available to all users free of cost [14]. GROMOS96 is a united atom force field, explicitly modeling only polar hydrogen atoms. Aliphatic hydrogens are coupled with attached carbons to form united atoms and handled as single atoms. Such a minor reduction in modelling precision significantly saves computational time. GROMOS96 is built into the GROMACS molecular dynamics simulation package alongside other force fields, and it offers the option to modify their respective residue topology and terminal files. It was developed specifically for proteins [15].

2 Materials

2.1 Instruments and Hardware

A system dedicated to perform simulation experiments (*see* **Note 1**).

2.2 Recommended Operating System

Linux. Using Virtual Machines like Cygwin makes it possible to run on Windows as well.

2.3 Molecular Dynamics Simulation Environment

GROMACS [14]. A molecular dynamics engine capable of preparation, production run, and analysis of molecular dynamics (*see* **Note 2**).

2.4 Visualization and Graphical Software

1. VMD [16] and UCSF Chimera [17]. Either software can visualize 3D structural coordinate data and coordinate trajectory files (*see* **Note 3**).
2. Xmgrace [18]. Graphing software for plotting analysis data.

3 Methods

3.1 Peptide Input File

High-resolution experimentally determined structure by X-ray crystallography or NMR in PDB format. PDB files of peptides will be the input file. Protein PDB files may be downloaded from RCSB Protein Databank. Alternatively, structures can be predicted using homology modelling and de novo structure prediction tools, validated using the Ramachandran plot.

3.2 Experiments to Generate Peptide Conformational Landscape in Water

1. *pdb2gmx* program may be used to convert the input structure file to the GROMACS structure file (GRO) format along with a topology file that describes the atoms, bonds, pairs, angles, and dihedrals based on the chosen force field's residue topology (we recommend GROMOS96 53a6 force field) and protonation state of polar sidechains and the N- and C-termini. In case a regular topology is unavailable for a particular entry, *ab initio* calculations can be used to create one. A file is also created that defines positional restraints on heavy atoms in the input PDB file (*see* **Note 4**).
2. Using the *editconf* program, create a solvent box centered on the solute by default, detailing the box type and its size in terms of distance from the solute of interest. This program modifies the input GRO file to contain the box type-dependent dimensions.
3. Solvate this peptide-containing box with explicit solvent molecules using the *solvate* function. This will generate a GRO file and update the topology file, including the solvent molecules.
4. Neutralize the system using the *genion* function with a TPR file assembled using the *grompp* function and an appropriate molecular dynamics parameter MDP file. The neutralization process adds cations and anions to maintain a neutral charge, usually Na^+ and Cl^-.
5. Minimize the solvated charge-neutral system using the *mdrun* function with a TPR file created using the *grompp* function and a MDP file. Minimization relaxes the system and eliminates any steric clashes or faulty geometry. Typically, the method used for energy minimization is called the steepest descent method. The system's status can be verified with the change in its potential energy against time using the g_*energy* program.
6. Perform positional restrained molecular dynamics simulation by restraining the heavy atoms of the molecule using posre.itp file, thus allowing the solvent and solute to equilibrate around the simulating molecule. The function used is *mdrun*, with a TPR file generated via the *grompp* function and an appropriate MDP file with defined restraint groups. A simulation time of

1 ns with the constant NVT conditions and 1 ns with NPT conditions is desirable.

7. Execute the production run with NPT conditions without position restraints using the *mdrun* command.

3.3 Experiments with a Protein–Peptide Complex

1. Dock the peptide to a protein complex using appropriate docking software (*see* **Note 5**).
2. Obtain the best pose for the protein–peptide complex and treat it as a single system (standard peptide in water) with different index values for the peptide and the protein backbone atoms using the *make_ndx program*.

3.4 Peptide Membrane Interactions

1. Using the peptide PDB file, apply the *pdb2gmx* program to convert the input structure file to the GROMACS structure file (GRO) described in the standard water peptide.
2. Build the lipid membrane coordinate file using the MemBuilder server 2 [19], and download the forcefield parameters of gromos53a6_lipid.ff described by Kukol A [20].
3. Using the *trjconv* program, remove the periodicity from the lipid bilayer configuration and from the GRO file, and obtain the system's unit cell size and vectors.
4. Align the peptides appropriately and generate a box around the peptides using the *editconf* function based on the box dimensions noted above, merge the coordinate files of the lipids and peptides and update the topology file of the system to reflect the inclusion of the lipid bilayer.
5. Solvate this peptide-containing box with explicit solvent molecules using the *solvate* function. This will generate a GRO file and update the topology file to include the solvent molecules.
6. Neutralize the system using the *genion* program with a TPR file assembled using the *grompp* program and an appropriate MDP file. Minimize the solvated charge-neutral system using the *mdrun* program with a TPR file assembled using the *grompp* program and a MDP file.
7. Make an index entry using *make_ndx* such that protein and lipid groups are merged. Perform positional restrained molecular dynamics simulations, restraining the heavy atoms of the system using posre.itp to allow the solvent and solute to equilibrate without additional structural changes. The program used is *mdrun*, with a TPR file generated by running the *grompp* program with an appropriate MDP file with defined restrained groups. Simulate for 10 ns with the constant NVT conditions and 10 ns with NPT conditions.
8. Execute the production run with NPT conditions without position restraints using the *mdrun* program.

3.5 Key Programs of GROMACS Suite for Conformational Analysis

1. Using the *energy* program, plot the various properties available to determine the time evolution of conformations.
2. Using the *gyrate* function, plot the radius of gyration of the molecule over simulation time.
3. Using the *rmsf* function, plot the root mean square fluctuation (RMSF) over simulation time.
4. Using the *rms* function, plot the root mean square deviation (RMSD) over simulation time.
5. Using the *hbond* function, plot the changes in hydrogen bonds.
6. Using the *cluster* function, plot the number of clusters, the population of each cluster, and the most stable microstate.

4 Notes

1. Although the system requirements for GROMACS are low and remain consistent across versions, CUDA-enabled GPU is ideal for larger simulations with disk space of more than 100 GB.
2. File formats, algorithms, inputs, and simulation optimization may vary between versions of GROMACS.
3. Other visualization software, such as Pymol [20], may also be used. However, the advantage of VMD and UCSF chimera is cross operating system support, free licensing, and intuitive graphical user interface (GUI).
4. Invoke *pdb2gmx* by typing ***gmx pdb2gmx*** in the terminal. For more information about the input and output files, type *man gmx pdb2gmx*. This is applicable to all GROMACS programs.
5. Autodock AGFR 1.2 [21] toolkit and HPEPDOCK [22] web server is recommended for protein-peptide docking due to it being free of cost to use, optimization to protein–peptide docking and benchmarking.

References

1. Redkar A, Ramakrishnan V (2023) Chapter 2 - Modeling and simulation of peptides. In: Ramakrishnan V, Patel K, Goyal R (eds) De novo peptide design. Academic Press, London, pp 35–56
2. Ramakrishnan V, Ranbhor R, Durani S (2004) Existence of specific "folds" in polyproline II ensembles of an "unfolded" alanine peptide detected by molecular dynamics. J Am Chem Soc 126:16332–16333
3. Ramakrishnan V, Ranbhor R, Durani S (2005) Simulated folding in polypeptides of diversified molecular tacticity: implications for protein folding and de novo design. Biopolymers 78: 96–105
4. Ramakrishnan V, Ranbhor R, Kumar A et al (2006) The link between sequence and conformation in protein structures appears to be stereochemically established. J Phys Chem B 110:9314–9323
5. Ranbhor R, Ramakrishnan V, Kumar A et al (2006) The interplay of sequence and stereochemistry in defining conformation in proteins and polypeptides. Biopolymers 83:537–545

6. Hazam PK, Jerath G, Kumar A et al (2017) Effect of tacticity-derived topological constraints in bactericidal peptides. Biochim Biophys Acta Biomembr 1859:1388–1395
7. Hazam PK, Jerath G, Chaudhary N et al (2018) Peptido-mimetic approach in the design of syndiotactic antimicrobial peptides. Int J Pept Res Ther 24:299–307
8. Hazam PK, Akhil R, Jerath G et al (2019) Topological effects on the designability and bactericidal potency of antimicrobial peptides. Biophys Chem 248:1–8
9. Hazam PK, Phukan C, Akhil R et al (2021) Antimicrobial effects of syndiotactic polypeptides. Sci Rep 11:1823
10. Goyal R, Jerath G, Chandrasekharan A et al (2020) Peptide-based delivery vectors with pre-defined geometrical locks. RSC Med Chem 11:1303–1313
11. Goyal R, Jerath G, Chandrasekharan A et al (2022) Molecular hybridization combining tumor homing and penetrating peptide domains for cellular targeting. Drug Deliv Transl Res 12:1285–1292
12. Jerath G, Goyal R, Trivedi V et al (2019) Syndiotactic peptides for targeted delivery. Acta Biomater 87:130–139
13. Jerath G, Darvin P, Christian Y et al (2022) Delivery of small molecules by syndiotactic peptides for breast cancer therapy. Mol Pharm 19:2877–2887
14. Van Der Spoel D, Lindahl E, Hess B et al (2005) GROMACS: fast, flexible, and free. J Comput Chem 26:1701–1718
15. Bonvin AMJJ, Mark AE, van Gunsteren WF (2000) The GROMOS96 benchmarks for molecular simulation. Comput Phys Commun 128:550–557
16. Humphrey W, Dalke A, Schulten K (1996) VMD: visual molecular dynamics. J Mol Graph 14:33–38
17. Pettersen EF, Goddard TD, Huang CC et al (2004) UCSF Chimera—a visualization system for exploratory research and analysis. J Comput Chem 25:1605–1612
18. Turner P (2005) XMGRACE, Version 5.1. 19. Center for Coastal and Land-Margin Research, Oregon Graduate Institute of Science and Technology, Beaverton, OR
19. Ghahremanpour MM, Arab SS, Aghazadeh SB et al (2014) MemBuilder: a web-based graphical interface to build heterogeneously mixed membrane bilayers for the GROMACS biomolecular simulation program. Bioinformatics 30: 439–441
20. Kukol A (2009) Lipid models for united-atom molecular dynamics simulations of proteins. J Chem Theory Comput 5:615–626
21. Zhang Y, Forli S, Omelchenko A et al (2019) AutoGridFR: improvements on AutoDock affinity maps and associated software tools. J Comput Chem 40:2882–2886
22. Zhou P, Jin B, Li H et al (2018) HPEPDOCK: a web server for blind peptide–protein docking based on a hierarchical algorithm. Nucleic Acids Res 46:W443–W450

Chapter 4

Peptide Synthesis

Debika Datta and Gaurav Pandey

Abstract

Peptide synthesis is characterized by forming a substituted amide linkage between two amino acids. The first successful attempt to synthetically link amino acids through peptide bonds is more than a hundred years old. However, the process and chemistry have undergone a series of optimizations, and peptide synthesis has become a more practical part of present-day high-throughput scientific research following the advent of solid-phase synthesis strategies. This article describes a generalized protocol for solid-phase peptide synthesis (SPPS) using Fmoc chemistry, including their purification and characterization by reversed-phase high-performance liquid chromatography and mass spectrometry.

Key words Peptide synthesis, Solid phase peptide synthesis, Fmoc chemistry, MALDI, HPLC

1 Introduction

Polymers of amino acids with roughly 2–50 amino acids are generally referred to as peptides [1]. Peptides have emerged as increasingly important biological molecules and have been employed in a myriad of applications, ranging from anti-microbial [2–5], anti-aggregation agents [6] to drug delivery [7–11]. Solid phase peptide synthesis (SPPS), developed by Robert Bruce Merrifield, revolutionized the process of peptide synthesis [12]. Merrifield was awarded the Nobel Prize in Chemistry in 1984, for this invention.

In the SPPS, a higher molar equivalent of reagents, including the amino acids, is employed. The application of surplus reagents aims to increase the probability of substituting the resin. At the same time, the unreacted fraction is washed off between subsequent reaction steps to avert undesirable reactions and reduce the formation of unwanted by-products. In SPPS, the sequence of synthesis proceeds from the C → N terminus, which is contrasting when compared to the natural protein biosynthesis (N → C direction). The peptide chain is linked with the resin at the C terminus. The N terminus of the following amino acid is coupled to the carboxyl group of the previous amino acid through amide bond formation,

Vibin Ramakrishnan (ed.), *Biophysical Characterization of Functional Peptides*, Springer Protocols Handbooks,
https://doi.org/10.1007/978-1-0716-3405-9_4,

and it is extended to full length from C terminus to N terminus. The basic solid-phase synthesis approach involves the covalent binding of building blocks comprising of a protective group and solid support material, also known as the resin. The protecting group (Table 1) protects the reactive functional group of the building block [1]. The synthesis is done sequentially, in a single vial, involving repetitive cleavage, activation, and coupling steps, the culmination of the sequence is followed by chemical cleavage of the final sequence from the solid support. Being more efficient and economical than liquid-phase peptide synthesis (LPPS), the SPPS offers numerous advantages, including higher purity, convenient purification, facilitating longer sequence synthesis, scalability, and compatibility with automated synthesis platforms [13].

The protocol below describes step-by-step synthesis of a peptide sequence present in the core motif of the β-amyloid peptide $A\beta_{16-22}$ [14]. The sequence synthesized is Ac-KLVFFAE-am, which has end capping in the form of N-terminal acetylation and C-terminal amidation.

Table 1
List of common protecting groups and the reagents used for the de-protection

Name and chemical structure	Cleavage reagent
Fmoc (9-Fluorenylmethoxycarbonyl) 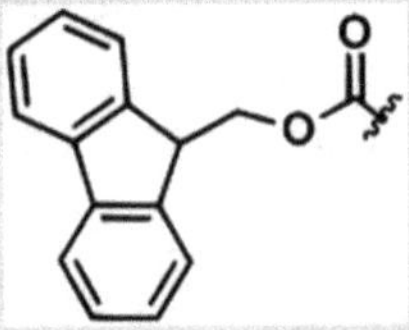	Piperidine (20%) in dimethyl formamide
Boc (tert-butyloxycarbonyl) 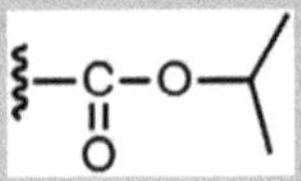	Trifluoroacetic acid
Trt (Trityl) 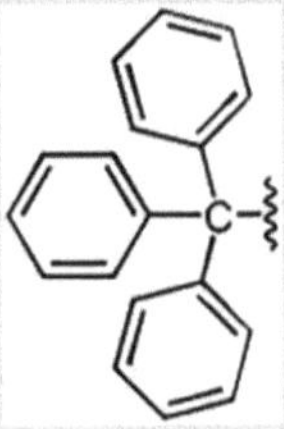	Trifluoroacetic acid

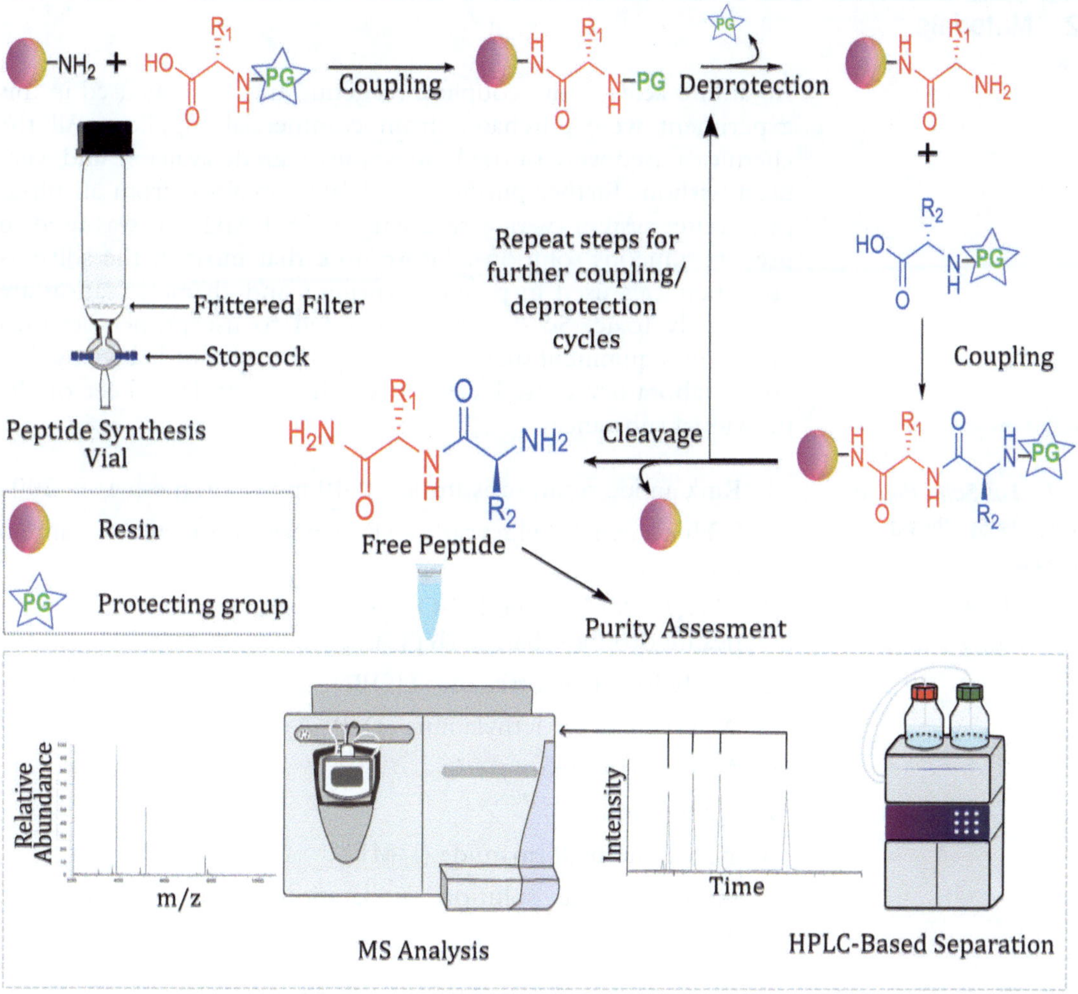

Fig. 1 Schematic overview of SPPS and characterization

The protocol includes five sections:

1. Manual synthesis of C-terminal amidated peptides,
2. N-terminal acetylation reaction,
3. Cleavage and side chain de-protection of the peptide from the resin,
4. Purity check of the synthesized peptide in a reversed-phase high-performance liquid chromatography (RP-HPLC) system, and
5. Mass analysis using matrix-assisted laser desorption/ionization-time of flight (MALDI-TOF) mass spectrometry. The protocol is very easy to execute and can be utilized to synthesize any end capped peptide for different applications in the lab. Figure 1 shows the schematic representation of the SPPS workflow.

2 Materials

All amino acids, resin, coupling reagents, and solvents used in this experiment were purchased from commercial suppliers. All the chemicals used were of the highest purity grade available and were used without further purification. Water obtained from an ultrapure water system (with a resistivity of 18.1 MΩ cm) was used to prepare aqueous solutions. Please note that most of the solvents and chemicals used in peptide synthesis and characterization are potentially toxic. So it is recommended to use proper personal protective equipment such as gloves, splash-resistant safety goggles, and a laboratory coat. Please check the safety data sheet of the individual substances.

2.1 For Solid Phase Peptide Synthesis (SPPS)

1. Rink amide resin, substitution 0.40 mmol/g, mesh 100–200.
2. 9-Fluorenylmethyloxycarbonyl (Fmoc)-protected L-amino acids (aa).
3. *N*,*N*,*N′*,*N′*-Tetramethyl-O-(1H-benzotriazol-1-yl)uronium hexafluorophosphate (HBTU).
4. *N*-Hydroxybenzotriazole (HOBt).
5. *N*,*N*-Diisopropylethylamine (DIPEA).
6. Acetic anhydride.
7. Piperidine.
8. *N*,*N*-dimethylformamide (DMF).
9. Fmoc removal solution preparation: 20% v/v Piperidine in DMF.

2.2 For Cleavage and Side Chain De-protection of Peptide from the Resin

1. Cleavage cocktail preparation: Trifluoroacetic acid (TFA), *m*-cresol, thioanisole, ethanedithiol (EDT) (20:2:2:1% v/v/v/v).
2. Diethyl ether.

2.3 For Purification by High-Performance Liquid Chromatography (HPLC) and Mass Analysis

1. HPLC Grade acetonitrile.
2. Solvent A preparation: Acetonitrile/Milli-Q water/TFA (10: 90:0.1% v/v/v).
3. Solvent B preparation: Acetonitrile/Milli-Q water/TFA (90: 10:0.1% v/v/v).
4. α-cyano-4-hydroxycinnamic acid (HCCA).
5. Solvent C preparation: Acetonitrile/Milli-Q water/TFA (30: 70:0.1% v/v/v).

2.4 Equipment

1. Peptide synthesis glass vessel (20–25 mL volume).
2. Nitrogen gas purging source tank.
3. General lab equipment such as weighing balance, tube rotator, centrifuge, vacuum concentrator, vortex, water bath sonicator, pH indicator strips, microcentrifuge tubes, pipette, ice bath, and dry ice.
4. Analytical reversed-phase high-performance liquid chromatography (RP-HPLC) system equipped with a C18 column.
5. MALDI-TOF mass spectrometer.
6. Clean lab glassware.
7. −80 °C Freezer/liquid nitrogen.

3 Methods

3.1 Manual Synthesis of C-terminal Amidated Peptides

1. Overnight resin swelling: Weigh 0.10 mmol of the Rink amide resin in a clean glassware and transfer the resin to a peptide synthesis glass vial. Soak the resin overnight in ~10 mL of DMF to ensure swelling. Mix the resin in the DMF solution a few times manually with a pipette. Alternatively, the vial can also be attached to a tube rotator for uniform mixing (*see* **Note 1**).
2. On the day of synthesis, first remove the DMF solution by filtration through the fritted glass using vacuum suction from below.
3. Fmoc removal from the resin: Perform this step only if the resin has a Fmoc group attached to it. Add 5 mL of Fmoc removal solution (20% v/v piperidine in DMF) with a glass pipette and keep the vial for mixing in the tube rotator for 20 minutes (min) (*see* **Note 2**).
4. Remove the synthesis vial and flush out the solution. Add 10 mL of DMF with a pipette and wash the resin thoroughly by mixing it couple of times with the pipette. Flush out the DMF. Repeat the washing step at least 10 times till a neutral pH is achieved. To check the pH, take 5 μL of the solution from the vial and drop cast on a pH indicator strip and verify if the pH of the solution corresponds to 7. This step is very critical (*see* **Note 3**).
5. Weigh 3 equivalents (3 × 0.10 mmol) of aa, HOBt, and HBTU and 6 equivalents (6 × 0.10 mmol) of DIPEA for preparing the coupling mixture (*see* **Note 4**).
6. Pre-activation of the coupling mixture: Add the weighed aa, HOBt, HBTU, and DIEPA in 5 mL of DMF in a microcentrifuge tube and vortex to dissolve the components completely. Set in the rotator for 10 min for mixing (*see* **Note 5**).

7. First coupling reaction: Drain the residual DMF from the resin from **step 4** and add the pre-activated coupling mixture from **step 6** onto the resin. It is optional to add a little extra (~1 mL) DMF to ensure the resin is free flowing in the solution. Purge nitrogen gas for 3–5 min for moisture removal. Set the synthesis vial in the tube rotator for mixing for 1 h (*see* **Note 6**).
8. Remove the synthesis vial and flush out the solution. Wash with 5 mL of DMF 5 times (5 × 5 mL).
9. Repeat **steps 5** and **6** for preparation and pre-activation of the coupling mixture for second coupling.
10. Second coupling reaction: Drain the residual DMF from the resin from **step 8** and add the pre-activated coupling mixture onto the resin. It is optional to add a little extra (~1 mL) DMF to ensure the resin is free flowing in the solution. Purge nitrogen gas for 3–5 min for moisture removal like previously. Set the synthesis vial in the tube rotator for mixing for 30 min.
11. Remove the synthesis vial and flush out the solution. Wash with 5 mL of DMF 5 times (5 × 5 mL).
12. Storage of resin in between synthesis: If done for the day, wash the resin with 5 mL diethyl ether 5 times (5 × 5 mL) and allow it to dry completely (*see* **Note 7**). Seal the cap of the synthesis vial with parafilm and store the resin for continuation later. Skip this step if you are not storing the resin and proceed to **step 14**.
13. Re-swelling of resin: If peptide synthesis is paused in between the completion of the sequence, on the day of resuming, add 10 mL of DMF to the resin and soak in the synthesis vial for 1 h before proceeding to **step 14**.
14. Fmoc removal from the amino acid: Add 5 mL of Fmoc removal solution (20% v/v piperidine in DMF) with a glass pipette and keep the vial for mixing in the tube rotator for 20 min.
15. Repeat **steps 4–14** till the attachment of the last aa from the sequence is complete.
16. After attaching the last aa and removing the Fmoc from the aa, wash the resin thoroughly with DMF (as per **step 4**) until a neutral pH is obtained. Proceed to Subheading 3.2 for end capping with the Acetyl group.

3.2 N-terminal Acetylation Reaction

1. Measure 5 equivalents (5 × 0.10 mmol) of acetic anhydride and DIPEA for preparing the coupling mixture. Add the required volume of both in 5 mL of DMF in a microcentrifuge tube and mix manually with a pipette. Add the activation mixture to the resin and set the vial in the rotor for coupling for 1 h (*see* **Note 8**).

2. Remove the vial from the rotor and wash the resin with 5 mL DMF 5 times (5 × 5 mL).
3. Repeat **steps 1** and **2** for one more time for acetylation.
4. Wash the resin with 5 mL diethyl ether 5 times (5 × 5 mL) and allow it to dry completely. Can store the resin-bound peptide after this step. Otherwise, proceed to Subheading 3.3 to remove the peptide from the resin.

3.3 Cleavage and Side Chain Deprotection of the Peptide from the Resin

NOTE: All procedures involving TFA must be performed in a fume hood. Please wear proper personal protection at all times. Please ensure the safe handling and disposal of the chemicals.

1. Add ~5 mL of the cleavage cocktail mixture (*m*-cresol, thioanisol, EDT, and TFA in a 2:2:1:20 ratio into the mixture) to the peptide-resin (*see* **Note 9**).
2. Incubate for a minimum of 12 h or a maximum of 15 h.
3. Prepare ice-cold ether by keeping the diethyl ether bottle in an ice bath/dry ice. Keep for at least 20 min for the ether to cool down.
4. Remove the resin by filtration and collect the peptide-containing solution in a vial. If required, wash the resin with a minimum volume of TFA.
5. For every 1 mL of the filtrate, prepare 10 mL of the ice-cold ether in separate 15 mL glass centrifuge tubes. Add 1 mL of the filtrate dropwise to the chilled ether to precipitate the peptide (*see* **Note 10**).
6. Cover the top of the tubes with aluminum foil and keep them in an ice bath inside the fume hood for 1 h.
7. Centrifuge the tubes containing the peptides for 10 min or till the peptide precipitates.
8. Decant the ether supernatant. Re-suspend the peptide in 2 mL of fresh ice-cold ether and vortex for 5 min.
9. Add 8 mL more of chilled ether to the suspension and repeat **steps 6–8** for, 8–10 times (*see* **Note 11**).
10. After the last centrifugation step, decant the ether supernatant and leave the peptide precipitated in the centrifuge tube. Make small perforations with a syringe needle in the aluminum foil and cover the tube. Keep the tube in the fume hood for a few hours to obtain the dried peptide and remove any residual TFA.
11. Alternatively, to lyophilize the peptide, dissolve it in ultrapure water, and freeze in a −80 °C freezer (overnight)/liquid nitrogen (instant). Place on the lyophilizer overnight.

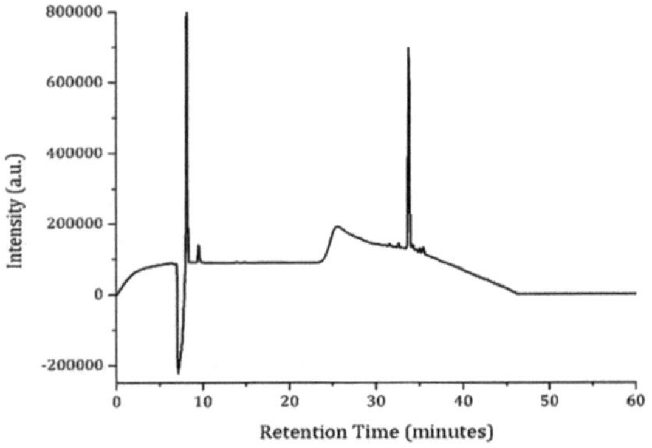

Fig. 2 HPLC chromatogram for the Ac-KLVFFAE-am peptide sequence. Peptide elution time is approximately 35 min

3.4 Purity Check of the Synthesized Peptide in a Shimadzu Prominence Modular HPLC System Equipped with a PDA Detector and a Reversed-Phase C18 Analytical Column

1. Dissolve ~1–2 mg of peptide in 1 mL ultrapure water. Run the solution in a linear gradient of acetonitrile solution by loading the sample using a syringe/autosampler vial (*see* **Note 12**).
2. Monitor the chromatogram at 210 nm for all amino acids. For aromatic aa containing peptides like Phe, Tyr, and Trp, monitor at 254 nm, 280 nm, and 280 nm, respectively. Figure 2 shows the HPLC chromatogram of the example peptide.
3. Collect the HPLC fraction corresponding to the peptide peak for mass analysis (*see* **Note 13**).

3.5 Mass Analysis Using MALDI-TOF Mass Spectrometry (Bruker AutoFlex Speed)

1. Prepare a saturated solution of HCCA matrix (~4 to 5 mg HCCA/mL) in solvent C (*see* **Note 14**).
2. Take 2 μL of the peptide HPLC fraction collected in the previous step and mix it with 2 μL of the HCCA matrix saturated solution (1:1 ratio matrix/analyte).
3. Deposit 0.5–1 μL of the mixture onto the MALDI plate and allow it to air dry. Run the sample in the MALDI-TOF instrument for mass analysis.
4. Compare the theoretical and expected mass. Consider the formation of ion adducts during analysis (Fig. 3).

4 Notes

1. Resin substitution and mesh size affect the reaction kinetics and yield of the peptide synthesized. Check with the manufacturer the substitution value of the resin before starting the synthesis. The swelling of the resin is critical in SPPS since the chemical reaction kinetics is controlled by the diffusion process. Therefore, a well-swelled resin has a higher rate of diffusion for the coupling reagents to reach the core and thereby reduces the

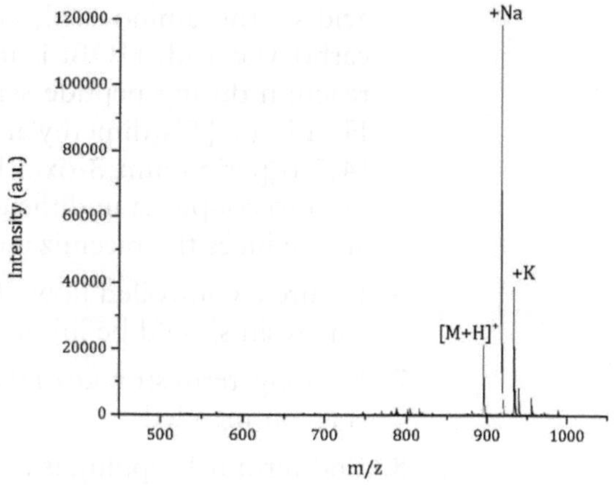

Fig. 3 MALDI-TOF mass spectrum of synthesized peptide: The first labelled peak denotes the observed mass $[M + H]^+$, while the other two indicate the mass with sodium $[M + Na]^+$, and potassium $[M + K]^+$ adducts, respectively

reaction times and enhances complete chemical conversions. Molecular sieves can be added to the DMF container to remove residual moisture.

2. Some of the commercial resins do not have a Fmoc protecting group attached. The attachment of the first amino acid could be done directly. Verify the chemical structure of the resin prior to the Fmoc removal step from the resin.
3. It is important to ensure complete de-protection and removal of Fmoc moiety in between the aa coupling steps to avoid side reactions. Multiple times washing with DMF is recommended till a neutral pH is obtained. To check the pH of an organic solvent using a pH indicator strip, add a drop of water followed by a drop of the solution to the strip. Occasionally mixing the resin with the pipette tip is required to prevent resin aggregation.
4. SPPS can be performed at any scale. An excess of aa and coupling reagents is generally added to ensure complete attachment of the aa. Calculations for activation mixtures: All calculations should be based on the scale of synthesis. In this case, the amount of resin taken is 0.1 mmol. Thus, for Fmoc-protected amino acids/HOBt/HBTU: 0.1 mmol × 3 × molecular weight of aa/HOBt/HBTU will give the mass of each of the chemicals to be used. For DIPEA: 0.1 mmol × 6 × molecular weight of DIPEA/density will provide the volume required for coupling. For acetic anhydride: 0.1 mmol × 5 × molecular weight of acetic anhydride/density will give the volume required for coupling.
5. Sometimes, slight coloration is observed for some of the aa activation. DIPEA or Hünig's base deprotonates the carboxylic

acid of the amino acid, and HBTU is used to activate free carboxylic acid. HOBt is used to avoid racemization as a side reaction during peptide synthesis. An alternative to HBTU is HATU (1-[Bis(dimethylamino)methylene]-1H-1,2,3-triazolo [4,5-b]pyridinium 3-oxid hexafluorophosphate) is sometimes used to couple aa in difficult sequences, as HATU reacts faster and reduces the racemization during coupling.

6. Ensure a controlled flow of the N_2 gas to avoid the resin spill. The resin should be immersed in the solution at all times.
7. For long-term storage of the resin, keeping it at −20 or −80 °C is recommended.
8. End terminal capping is a common modification in chemically synthesized peptides. N-terminal acetylation and C-terminal amidation remove the positive and negative charges, respectively, and also block from any unwanted side product and impurity formation.
9. TFA is a very strong fuming acid. Ensure proper handling of the chemical during synthesis. In Fmoc SPPS, a high percentage (>80%) of TFA is generally sufficient to cleave the peptide from the resin. Other scavengers are added to the cleavage cocktail solution to prevent the generated reactive cationic groups from reattaching to the peptide. Among the different cleavage mixtures, another combination: TFA/TIS (triisopropylsilane)/water (95:2.5:2.5) is also very efficient. Around ~10 to 30 mL of cleavage mixture per gram of resin can be used.
10. The ice-cold ether volume is ~8 to 10 times that of the drained filtrate.
11. The peptide can be washed and centrifuged with ether repeatedly to remove any residual scavengers.
12. 0.1% TFA is used as the ion pairing agent in HPLC solvent preparation. The following column was used in characterizing the example peptide. Column specifications: Oyster BDS premium C18 5 μm Dimension: 4.6 mm × 250 mm. Particle size: 5 μm Flow rate: 0.5 mL/min.
13. Once the mass is validated, if required, further purification of the peptide can be done using a semi-preparative C18 RP-HPLC column.
14. Saturated MALDI matrix solutions are often difficult to mix. Vortexing or water bath sonication expedites the process of dissolution. Once prepared should be stored at 4 °C. Other common types of matrices are 2,5-dihydroxybenzoic acid and Sinapinic acid.

References

1. Linder J (2012) The science behind peptides. Plast Surg Nurs 32(2):71–72
2. Hazam PK, Akhil R, Jerath G, Saikia J, Ramakrishnan V (2019) Topological effects on the designability and bactericidal potency of antimicrobial peptides. Biophys Chem 248:1–8
3. Hazam PK, Singh A, Chaudhary N, Ramakrishnan V (2019) Bactericidal potency and extended serum life of stereo-chemically engineered peptides against Mycobacterium. Int J Pept Res Ther 25(2):465–472
4. Hazam PK, Phukan C, Akhil R, Singh A, Ramakrishnan V (2021) Antimicrobial effects of syndiotactic polypeptides. Sci Rep 11(1): 1823
5. Hazam PK, Jerath G, Kumar A, Chaudhary N, Ramakrishnan V (2017) Effect of tacticity-derived topological constraints in bactericidal peptides. Biochim Biophys Acta Biomembr 1859(8):1388–1395
6. Pandey G, Morla S, Kumar S, Ramakrishnan V (2020) Modulation of tau protein aggregation using 'Trojan' sequences. Biochim Biophys Acta Gen Subj 1864(7):129569
7. Goyal R, Jerath G, Chandrasekharan A, Christian Y, Kumar TRS, Ramakrishnan V (2022) Molecular hybridization combining tumor homing and penetrating peptide domains for cellular targeting. Drug Deliv Transl Res 12(5):1285–1292
8. Jerath G, Goyal R, Trivedi V, Santhoshkumar TR, Ramakrishnan V (2019) Syndiotactic peptides for targeted delivery. Acta Biomater 87:130–139
9. Jerath G, Darvin P, Christian Y, Trivedi V, Kumar TRS, Ramakrishnan V (2022) Delivery of small molecules by syndiotactic peptides for breast cancer therapy. Mol Pharm 19(8): 2877–2887
10. Goyal R, Jerath G, Akhil R, Chandrasekharan A, Puppala ER, Ponneganti S, Sarma A, Naidu VGM, Santhoshkumar TR, Ramakrishnan V (2021) Geometry encoded functional programming of tumor homing peptides for targeted drug delivery. J Control Release 333:16–27
11. Goyal R, Jerath G, Chandrasekharan A, Kumar TRS, Ramakrishnan V (2020) Peptide-based delivery vectors with pre-defined geometrical locks. RSC Med Chem 11(11):1303–1313
12. Merrifield RB (1963) Solid phase peptide synthesis. I. The synthesis of a tetrapeptide. J Am Chem Soc 85(14):2149–2154
13. Isidro-Llobet A, Kenworthy MN, Mukherjee S, Kopach ME, Wegner K, Gallou F, Smith AG, Roschangar F (2019) Sustainability challenges in peptide synthesis and purification: from R&D to production. J Organic Chem 84(8):4615–4628
14. Pandey G, Morla S, Kumar S, Ramakrishnan V (2020) Modulating Aβ fibrillogenesis with 'Trojan' peptides. Neuropeptides 81:102030

Chapter 5

Mass Spectrometry

Ilackkeya Bhavananthi and Vibin Ramakrishnan

Abstract

Synthetic peptides have made promising advances in cancer treatment, drug delivery vehicles, antimicrobial drugs, and vaccine design. The highly pure peptides after synthesis and purification are characterized using matrix-assisted laser desorption ionization—time of flight mass spectrometry (MALDI-TOF MS). This highly sensitive instrument employs a soft ionization technique to characterize the mass of peptides and proteins. This chapter outlines the choice of matrix selection, the procedure for sample preparation, and the mass spectrometric characterization.

Key words Synthetic peptides, MALDI-TOF, Mass spectrometry, Peptide characterization, HCCA, MALDI matrix

1 Introduction

Mass spectrometry is one of the most effective techniques for characterizing peptides, where the mass-to-charge ratio (m/z) of ions is determined. The introduction of "soft ionization" techniques like matrix-assisted laser desorption ionization (MALDI) and electron spray ionization (ESI) allows rapid analysis of the mass of synthetic peptides [1]. Mass spectrometry is therefore an important analytical tool in the complete work flow of structural characterization of functional peptides [2–7].

Karas and Hillenkamp introduced MALDI in 1988, which works on the principle where a strong energy-absorbent, organic matrix absorbs the laser of a particular wavelength (337 nm) to desorb and ionize, thus generating singly protonated or deprotonated ions for the peptide under study. In the positive and negative ion mode of data acquisition, the organic matrix gets either protonated or deprotonated and produces $M - H^+$ or $M - H^-$ ion, respectively. The positive ion mode tends to produce multiple singly charged ions, including $M - Na^+$ and $M - K^+$ peaks in the spectra, whereas the negative ion mode detection reduces the fraction of other peaks and produces a highly resolved $M - H^-$ ion

Vibin Ramakrishnan (ed.), *Biophysical Characterization of Functional Peptides*, Springer Protocols Handbooks,
https://doi.org/10.1007/978-1-0716-3405-9_5,

peak. The negative ion mode has received little attention, due to which they are not used for detecting the mass of larger biomolecules. However, phosphorylated proteins and acidic peptides containing glutamic acid (E) and aspartic acid (D) tend to deprotonate rapidly and form an $M - H^-$ ion in the negative ion mode, which gives us evidence that negative ion mode analysis can complement and validate the protein and peptide analysis in positive ion mode [8–10].

MALDI being a discontinuous source of ion production pairs well with the TOF analyzers specifically for peptide mass analysis. The detector is used to measure the m/z ratio of an analyte of interest by calculating the time taking to reach the detector [11, 12]. Different mass ranges of analytes are detected in two different modes: linear-TOF mode and reflectron-TOF mode. The linear-TOF mode detects high molecular weight compounds, whereas the reflectron-TOF mode detects small organic molecules with low molecular weights at high resolution and accuracy [13]. This chapter provides a detailed protocol highlighting the matrix selection, sample preparation, and mass analysis of synthetic peptides in a MALDI-TOF mass spectrometry analyzer (Fig. 1).

2 Materials

1. Peptide or protein of interest in a suitable solvent based on the choice of matrix.
2. Matrix solution preparation (*see* Subheading 3.3).
3. Solvents—Acetonitrile (HPLC grade), trifluoroacetic acid (TFA), ethanol, ultrapure water.
4. Solvent preparation.

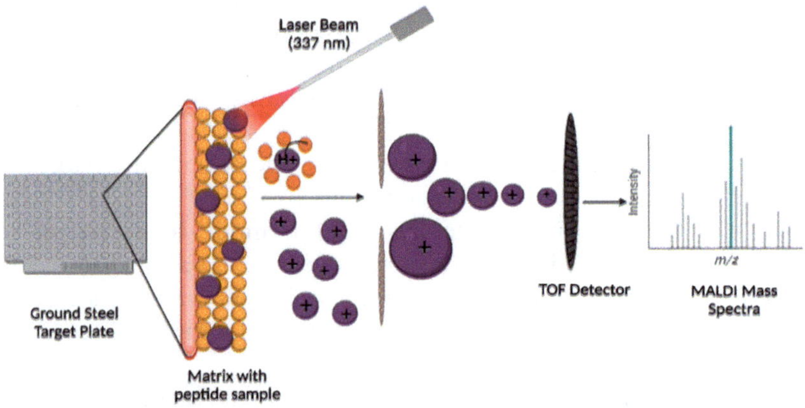

Fig. 1 Schematic diagram of MALDI-TOF MS analyzer. (Created with Biorender.com)

0.1% TFA—Dissolve 1 mL of TFA in 1000 mL of water.

AT30—Dissolve 30 mL of acetonitrile in 70 mL of 0.1% TFA.

AT50—Dissolve 50 mL of acetonitrile in 50 mL of 0.1% TFA.

AT85—Dissolve 85 mL of acetonitrile in 15 mL of 0.1% TFA.

AT90—Dissolve 90 mL of acetonitrile in 10 mL of 0.1% TFA.

AT95—Dissolve 95 mL of acetonitrile in 5 mL of 0.1% TFA.

5. Instrument—Bruker's Daltonics autoflex speed MALDI-TOF mass spectrometer (Fig. 2).

Fig. 2 Autoflex Speed MALDI-TOF mass spectrometer. (Bruker's Daltonics)

3 Methods

3.1 MALDI Target Cleaning

MALDI target plate has many wells/spots that hold the sample for analysis. The target plates are of different types: Ground Steel, AnchorChip (800 μm), BigAnchor (2000 μm), and AnchorChip Small (400 μm). The proper cleaning of the target plates is essential for sample analysis.

1. Prepare 2-propanol, AT30 solution, and fetch Milli-Q water.
2. Wipe off old samples in the target plates with 2-propanol wetted tissue followed by water.
3. Sonicate the target plate with 2-propanol and AT30 solution each for 10 min duration.
4. Dry the target plate with pure air or N_2. Do not wipe the front side of the target plate after it is cleaned (*see* **Note 1**).

3.2 Matrix Selection for Different Samples

Organic matrices of different types help in identifying and characterizing different peptides or protein samples of our interest. The choice of matrix selection depends on the sample of interest and the target plate available for MALDI-TOF analysis. Table 1 enlists different types of matrices used for peptide characterization.

3.3 Sample Preparation

Sample preparation is the pivotal step in the MALDI-TOF MS analysis. The peptide should get incorporated into the organic matrix such that the matrix absorbs the laser and gets desorbed and modulates the ionized peptide [14]. The sample and matrix preparation, as well as the sample spotting protocol, has been described below for all the types of matrices (Table 1). Depending on the peptide or protein of interest, the corresponding matrix, and the target type, the user can prepare their sample for mass analysis.

3.3.1 α-Cyano-4-Hydroxy Cinnamic Acid (HCCA)

3.3.1.1 Ground Steel

1. Prepare the sample in 0.1% TFA or AT30 solution.
2. Saturate the HCCA matrix in AT30 solution.
3. Mix equal volume (2 μL) of the matrix and sample solution.
4. Add 0.5 μL of pre-mixed solution into the target plate.

3.3.1.2 AnchorChip (800 μm)

1. Prepare the sample in 0.1% TFA or AT30 solution.
2. Dissolve 1.4 mg/mL HCCA in a solvent of 85:15 ACN: H_2O, 1 mM $NH_4H_2PO_4$, and 0.1% TFA.
3. Add a spot of 0.5–1 μL of the sample into the target plate.
4. Allow the sample to dry and add 0.5 μL of the matrix solution and dry it at normal conditions (*see* **Note 2**).
5. **External calibrant preparation**: Prepare "Peptide Calibrant Standard II" in 125 μL of AT30.

Table 1
Different types of matrix used to characterize different sample types

Matrix (abbreviation)	Sample	Target type
α-Cyano-4-hydroxy cinnamic acid (HCCA)	Peptides, protein digests (<10 kDa)	Ground Steel
		AnchorChip (800 μm)
	nanoLC-MALDI analysis of peptide mixtures	AnchorChip (800 μm)
2,5-Dihydroxy benzoic acid (2,5-DHB)	Peptides, glycoprotein digests, phosphoprotein digests, glycans, and intact proteins (>10 kDa)	Ground Steel
	Peptides, glycans, and glycoprotein digests	AnchorChip (800 μm)
2,5-Dihydroxy acetophenone (2,5-DHAP)	Intact proteins (>10 kDa)	Ground steel; BigAnchor (2000 μm)
A mixture of 2,5-DHB and 2-Hydroxy-5-methoxy benzoic acid (sDHB) (90:10)	Intact proteins (>10 kDa)	Ground steel
		BigAnchor (2000 μm)
		AnchorChip (800 μm)
Trans-3,5-dimethoxy-4-hydroxy cinnamic acid (SA)	Intact proteins (>10 kDa)	Ground steel
3-Hydroxy picolinic acid (3-HPA)	Oligonucleotides	Ground steel
		AnchorChip (800 μm)
		AnchorChip Small (400 μm)

6. Prepare the calibrant solution and HCCA matrix solution in 1: 200 ratio.
7. Add 0.5 μL of this mixture on to the neighboring spot of the sample position.

3.3.1.3 nanoLC-MALDI Analysis in AnchorChip (800 μm)

1. Prepare the sample in 0.1% TFA or AT30 solution.
2. Prepare the HCCA matrix solution (800 μL) by adding 36 μL of HCCA stock solution (saturated in AT90), 748 μL AT95, 8 μL of 100 mM $NH_4H_2PO_4$ (dissolved in water), and 8 μL of 10% TFA in water.
3. Spot approx. 420 mL of matrix solution using the syringe pump by setting the flow rate to 100 μL/h for 15 s spotting and 150 μL/h for 10 s spotting.

4. **External calibrant preparation**: Prepare "Peptide Calibrant Standard II" in 125 μL of AT30.
5. Mix 1.5 μL of calibration solution with 300 μL of the matrix solution. Spot 420 nL of this mixture on the neighboring spot of the sample position.

3.3.2 2,5-Dihydroxy Benzoic Acid (2,5-DHB)

3.3.2.1 Ground Steel

1. Prepare the sample in 0.1% TFA or AT30 solution. Use water for glycan samples.
2. Prepare the matrix solution by dissolving 20 mg/mL of 2,5-DHB in AT30.
   Additives (Optional): Add 1% H_3PO_4 for phosphopeptide analyses and add 1 mM NaCl, in case of glycan analyses.
3. Mix equal volume (2 μL) of the matrix and sample solution.
4. Add 0.5 μL mixture onto the target plate.

3.3.2.2 AnchorChip (800 μm)

1. Prepare the sample in 0.1% TFA or AT30 solution. Use water for glycan samples.
2. Prepare the matrix solution by dissolving 10 mg/mL of 2,5-DHB in AT30.
   Additives (Optional): Add 1 mM NaCl for glycan analyses.
3. Add 0.5 μL of the matrix solution on the target and let the matrix dry.
4. Spot 0.5–1 μL of the sample on top of the matrix and allow it to dehydrate at normal conditions.

3.3.3 2,5-Dihydroxy Acetophenone (2,5-DHAP)

3.3.3.1 Ground Steel and BigAnchor (2000 μm)

1. Prepare the sample in 0.1% TFA solution.
2. Prepare the matrix solution by dissolving 7.6 mg of 2,5-DHAP in 375 μL of 100% ethanol.
3. Add 125 μL of 18 mg/mL diammonium hydrogen citrate (DAC).
4. Deposit 2 μL of the sample in 2% of TFA solution and then add 2 μL of the matrix to it.
5. Pipette up and down until it crystallizes.
6. Spot 0.5 μL of the dried sample-matrix mixture onto the target plate.

3.3.4 Mixture of 2,5-DHB and 2-Hydroxy-5-Methoxy Benzoic Acid (sDHB) (90:10)

3.3.4.1 Ground Steel

1. Prepare the sample in 0.1% TFA or AT30 solution.
2. Prepare the matrix solution by dissolving 50 mg/mL sDHB in AT50.
3. Mix equal volume (2 μL) of the matrix and sample solution.
4. Add 0.5 μL mixture onto the target plate.

3.3.4.2 BigAnchor (2000 μm)

1. Prepare the sample in 0.1% TFA or AT30 solution.
2. Prepare the matrix solution by dissolving 25 mg/mL sDHB in AT50.
3. Spot 1 μL of the sample and let it dry.
4. Then add 1 μL of the matrix solution and allow it to dry at room temperature.

3.3.4.3 AnchorChip (800 μm)

1. Prepare the sample in 0.1% TFA or AT30 solution.
2. Prepare the matrix solution by adding 10 mg/mL sDHB in AT50.
3. Spot 0.5–1 μL of the sample and allow it to dry.
4. Then add 0.5 μL of the matrix to the dried sample spot and allow it to dry at normal conditions.

3.3.5 *Trans-3,5-Dimethoxy-4-Hydroxy Cinnamic Acid (SA)*

3.3.5.1 Ground Steel

1. Prepare the sample in 0.1% TFA solution.
2. Prepare the matrix solution I by saturating SA in ethanol.
3. Prepare the matrix solution II by saturating SA in AT30.
4. Prepare a thin matrix layer on the target using solution I.
5. Mix equal volume (2 μL) of the matrix solution II and sample solution.
6. Add 0.5 μL of the mixture onto the thin layer prepared before.

3.3.6 *3-Hydroxy Picolinic Acid (3-HPA)*

3.3.6.1 Ground Steel

1. Prepare the sample in water.
2. Prepare the matrix solution by saturating 3-HPA in 50:50 (v/v) ACN: H_2O containing 10 mg/mL DAC.
3. Add 0.5 μL of the matrix on the target and let it dehydrate.
4. Spot 0.5 μL of the sample and allow it to dry at normal conditions.

3.3.6.2 AnchorChip (800 μm)

1. Prepare the sample in water.
2. Make the matrix solution by dissolving half-saturated 3-HPA in 50:50 (v/v) ACN: H_2O and 10 mg/mL DAC.
3. Add 0.5 μL of the matrix solution on the target and let it dry.
4. Spot 0.5 μL of the sample and allow it to dry at normal conditions.

3.3.6.3 AnchorChip Small (400 μm)

1. Prepare the sample in water.
2. Make the matrix solution by adding 10 mg/mL 3-HPA in water and 1 mg/mL DAC.
3. Add 1 μL of matrix onto the target and let it dry.
4. Spot 1 μL of the sample onto the target plate and let it dry at normal conditions.

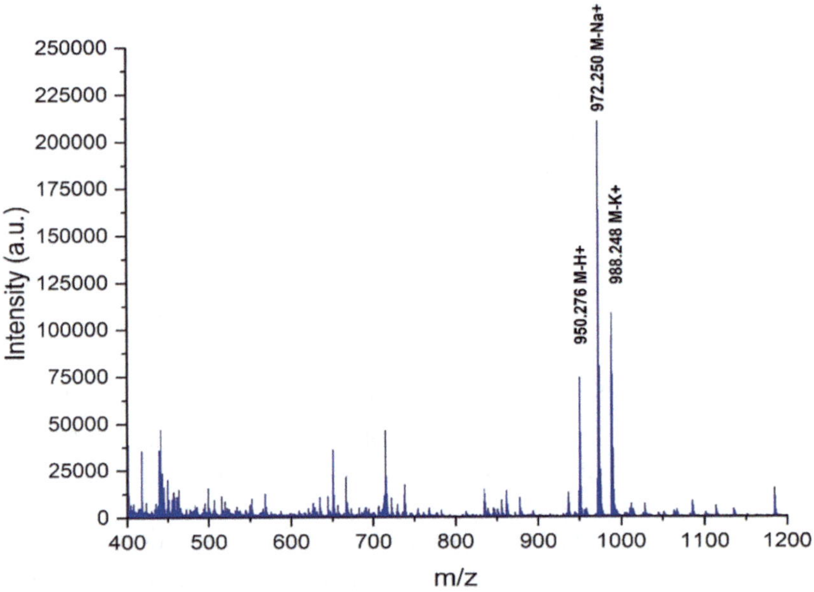

Fig. 3 Mass spectrum of the peptide sequence Ac-IHIHIYI-$CONH_2$ in positive ion mode

3.4 Mass Analysis

1. Once the sample is prepared and spotted onto the target plate, analyze the sample in MALDI-TOF analyzer integrated with the N_2 laser at a wavelength of 337 nm in positive ion mode [15, 16] (*see* **Note 3**).
2. Depending on the anticipated mass range of the sample for analysis, the linear mode or reflectron mode can be set.
3. Use optimum laser power to obtain an accurate signal and avoid burning the sample.
4. Obtain five consecutive laser shots of the sample, and the average of the data is obtained in the mass spectra.
5. Obtain the peptide mass fingerprint (PMF) and monoisotopic masses of the peptide from the spectrum as shown in Fig. 3 for the peptide sequence **Ac-IHIHIYI-$CONH_2$** [17] with the molecular weight of 949.15 Da (*see* **Note 4 and 5**).

4 Notes

1. Allow the plate to dry at room temperature, in the case of non-availability of pure gases.
2. Ensure that the sample is free of contamination to prevent interference with MALDI. In the presence of contaminants like alkaline salts or surfactants, clean the sample using magnetic beads, and dialysis membrane before submitting it for mass analysis.

3. To confirm that no matrix interferes with your peptide during analysis, it is always advised to keep a blank of matrix solution for analysis.
4. The presence of salt in the matrix solution or the sample will result in Na^+ (23 a.u.) and K^+ (39.1 a.u.) adducts in the spectra. The resulting spectra will have M − Na^+ and M − K^+ peaks along with the monoisotopic mass peak [14].
5. The matrix solution and sample preparation methods can produce varying results in complex peptide mixtures, due to ion suppression effects and discrimination in mass detection. To account for this, spectra can be obtained from different matrices [18].

References

1. Butler JM, Jiang-Baucom P, Huang M et al (1996) Peptide nucleic acid characterization by MALDI-TOF mass spectrometry. Anal Chem 68:3283–3287
2. Goyal R, Jerath G, Chandrasekharan A et al (2022) Molecular hybridization combining tumor homing and penetrating peptide domains for cellular targeting. Drug Deliv Transl Res 12:1285–1292
3. Saikia J, Dharmalingam K, Anandalakshmi R et al (2021) Electric field modulated peptide based hydrogel nanocatalysts. Soft Matter 17: 9725–9735
4. Goyal R, Jerath G, Akhil R et al (2021) Geometry encoded functional programming of tumor homing peptides for targeted drug delivery. J Control Release 333:16–27
5. Jerath G, Goyal R, Trivedi V et al (2020) Conformationally constrained peptides for drug delivery. Int J Mol Sci 26:e3244
6. Hazam PK, Akhil R, Jerath G et al (2019) Topological effects on the designability and bactericidal potency of antimicrobial peptides. Biophys Chem 248:1–8
7. Hazam PK, Jerath G, Chaudhary N et al (2018) Peptido-mimetic approach in the design of syndiotactic antimicrobial peptides. Int J Pept Res Ther 24:299–307
8. Bacher G, Körner R, Atrih A et al (2001) Negative and positive ion matrix-assisted laser desorption/ionization time-of-flight mass spectrometry and positive ion nano-electrospray ionization quadrupole ion trap mass spectrometry of peptidoglycan fragments isolated from various Bacillus species. J Mass Spectrom 36:124–139
9. Gao J, Cassady CJ (2008) Negative ion production from peptides and proteins by matrix-assisted laser desorption/ionization time-of-flight mass spectrometry. Rapid Commun Mass Spectrom 22:4066–4072
10. Mädler S, Barylyuk K, Boeri Erba E et al (2012) Compelling advantages of negative ion mode detection in high-mass MALDI-MS for homomeric protein complexes. J Am Soc Mass Spectrom 23:213–224
11. Chaurand P, Luetzenkirchen F, Spengler B (1999) Peptide and protein identification by matrix-assisted laser desorption ionization (MALDI) and MALDI-post-source decay time-of-flight mass spectrometry. J Am Soc Mass Spectrom 10:91–103
12. Singhal N, Kumar M, Kanaujia PK et al (2015) MALDI-TOF mass spectrometry: an emerging technology for microbial identification and diagnosis. Front Microbiol 6:791
13. Cho Y-T, Su H, Wu W-J et al (2015) Chapter Six - Biomarker characterization by MALDI–TOF/MS. In: Makowski GS (ed) Advances in clinical chemistry. Elsevier, Amsterdam, pp 209–254
14. Dave KA, Headlam MJ, Wallis TP et al (2011) Preparation and analysis of proteins and peptides using MALDI TOF/TOF mass spectrometry. Curr Protoc Prot Sci Chapter 16: 16.13.1–16.13.21
15. Gogichaeva NV, Alterman MA (2019) Amino acid analysis by means of MALDI TOF mass spectrometry or MALDI TOF/TOF tandem mass spectrometry. In: Alterman MA (ed) Amino acid analysis: methods and protocols. Springer, New York, NY, pp 17–31
16. Gowrishankar S, Kamaladevi A, Pandian SK (2022) MALDI-TOF analysis of actinobacterial peptides with respect to MASCOT database. In: Dharumadurai D (ed) Methods

in actinobacteriology. Springer, New York, NY, pp 273–283
17. Zozulia O, Dolan, Korendovych IV (2018) Catalytic peptide assemblies. Chem Soc Rev 47:3621
18. Cohen SL, Chait BT (1996) Influence of matrix solution conditions on the MALDI-MS analysis of peptides and proteins. Anal Chem 68:31–37

Chapter 6

Circular Dichroism (CD) Spectroscopy

Kalpana Kumari and Vibin Ramakrishnan

Abstract

Circular dichroism (CD) has been extensively used to characterize proteins, especially protein structural changes, protein–ligand interactions, and structural response to external stimuli. CD signals result from the interaction between chromophores that may possess intrinsic chirality or are kept in a chiral environment. In this chapter we discuss how this technique is used to determine the secondary structure of peptides.

Key words Polarize light, Differential absorption, Aromatic amino acid, Conformation, Chirality

1 Introduction

Circular dichroism is a technique that can be used to examine different aspects of optically active molecules. It provides structural information about the species present in the sample. CD measures differential absorption of left and right circularly polarized light.

$$\mathrm{CD} = A(L) - A(R) = \Delta\varepsilon Cl \tag{1}$$

Where absorbance (A) is related to the path length (l), protein concentration (C), and molar extinction coefficient (ε). The extent of absorption of left and right circularly polarized light varies for different samples. The side chains of tryptophan, tyrosine, phenylalanine, cystine are four intrinsic protein chromophores that absorb light in the near-UV region (340–255 nm) of the electromagnetic spectrum. Few amino acid residues, including tyrosine, tryptophan, phenylalanine, histidine, and methionine absorb light significantly in the far-UV region (<250 nm) [1]. The peptide bond is one of the most critical contributors here, with $n \rightarrow \pi^*$ and $\pi \rightarrow \pi^*$ transitions at wavelengths 220 nm and 190 nm, respectively. The far-UV CD bands show the secondary structure of a protein, including alpha-helix, beta-sheet, beta-turn, or random coil [2–4]. CD spectra of different secondary structure of a protein are

Vibin Ramakrishnan (ed.), *Biophysical Characterization of Functional Peptides*, Springer Protocols Handbooks,
https://doi.org/10.1007/978-1-0716-3405-9_6,

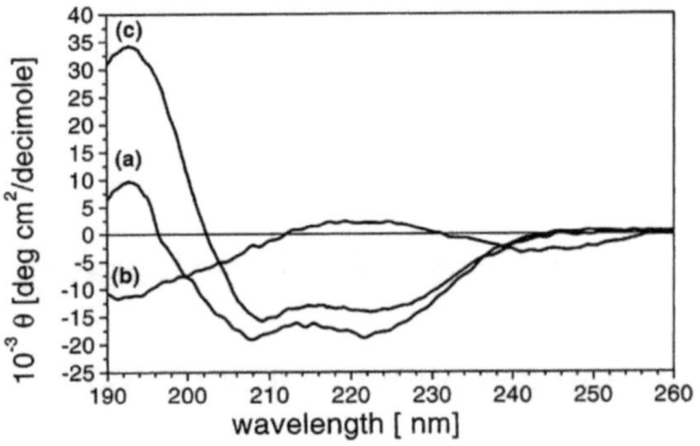

Fig. 1 CD spectra poly(L-arginine) retinoate having beta-sheet conformation (**a**), poly(L-histidine) retinoate having random coil conformation (**b**), and poly(L-lysine) retinoate having alpha-helical conformation (**c**) [5]. (Reprinted with permission from ACS Langmuir, 2000, 16:850–857. Copyright 2000 American Chemical Society)

shown in Fig.1. The molar extinction coefficients of tyrosine, tryptophan, cystine, and phenylalanine are given in Table 1.

Since CD spectra are very sensitive to molecular conformations, it is one of the widely used methods for protein analysis. It is susceptible to changes in the conformation irrespective of its origin and applies to many solvents for studying proteins [6].

CD spectroscopy can be used for several applications [7–19]. The most widely used are the following:

1. To estimate secondary structure content of protein [11–15].
2. To detect the conformational changes at different pH, presence of various salts, and changes in solvent [16–19].
3. To monitor protein denaturation at different temperatures or denaturants such as urea.
4. To monitor protein–ligand and protein–peptide interactions.
5. To study protein folding, unfolding, and refolding.

In this chapter, we present a standard protocol for the estimation of protein (or peptide) secondary structure.

2 Materials

1. CD instrument. Principal suppliers: Jasco Inc., Norlab, Scientex, Jasco UK, ATA Scientific, JASCO Global.
2. Quartz cuvettes with pathlength range: 0.1–10 mm.
3. For instrument calibration: Camphorsulfonic acid (1 mg/mL in a 1-mm cuvette).
4. Standard reagents and organic solvents should be of spectroscopic grade without any impurities.

Table 1
Extinction coefficient of amino acids at λ_{max}

Amino acid	Wavelength (λ_{abs}) (nm)	Extinction coefficient (M^{-1} cm^{-1})
Tyrosine	280	1280
Tryptophan	280	5690
Cystine	280	120
Phenylalanine	254	200

3 Methods

3.1 Instrument Preparation

1. It is a crucial step in CD experiment. All oxygen molecules should be removed from the instrument by purging pure nitrogen 15–20 min before switching on the lamp. Otherwise, it might form ozone that may damage the optical device.
2. CD instrument must be calibrated routinely by using camphorsulfonic acid.

3.2 Sample Preparation

1. Weigh the purified and lyophilized peptide and dissolve in the buffer of interest. The concentration of peptides must be accurate. It should not depend on the dry mass of the peptide. The concentration of stock peptide must be quantified by measuring absorbance with UV-Vis spectroscopy and calculated using Beer-Lambert law (*See* **Note 4** and **Note 5**)[20, 21]. Measure the absorbance at 280 nm, if the peptide contains tyrosine, tryptophan, and cystine. The extinction coefficient of the peptide is calculated as:

$$\varepsilon_{280} = x\varepsilon_{tyr} + y\varepsilon_{trp} + z\varepsilon_{cys} \quad (2)$$

   where x = Number of tyrosine residues.
   y = Number of tryptophan residues.
   z = Number of cystine residues.
   Absorbance should be in the range of 0.1–1.0 a.u. If absorbance is too low, use cuvette of longer path length, whereas if it is too high, use cuvette of shorter path length or dilute the sample (*See* **Note 1**).
2. To determine effect of pH, concentration of the peptide should remain same, while the concentration of buffer with different pH may vary.
3. The peptide concentration must be constant, while the salt of buffer concentration may vary for the study of ionic strength.
4. To study the concentration dependency, the condition of buffer remains constant, while the concentration of peptide may be different.

3.3 Data Collection

1. Place the sample in the holder and scan it from 280 to 180 nm. Other parameters, including slit width, time constant, and the average number of scans, should be uniform for a study. Try to take 5–8 scans per sample. These parameters can be saved as a method file, and same method file can be used for further study. The HT voltage must be below 600 V (*See* **Note 3**) [22, 23].
2. Record baseline scan of buffer (only) by using the same parameters.

3.4 Data Analysis

1. A baseline scan of the buffer must always be deducted from the sample (peptide) scan.
2. Data obtained from the CD is shown in millidegrees. It can also be expressed in molar ellipticity.

$$\theta = \theta_{\text{observed}}\left(\frac{\text{MRW}}{10\,cd}\right) \tag{3}$$

where θ_{observed} = Ellipticity (measured experimentally).

MRW = Mean residue molecular weight of the peptide (Molecular weight/Total number of peptide bonds).

c = Peptide concentration (mg/mL).

d = Pathlength of the cuvette (cm).

3. Several software such as DichroWeb, K2D2, CDtoolX, etc. may be used to determine the relative sequence length corresponding to different secondary structures such as α-helix, beta-sheet, or even random coil structures that cannot be categorized as a secondary structure (*See* **Note 2**) [24–26, 27].

4 Notes

1. Cuvettes should be properly cleaned after each use of a sample by using a cleaning solution such as chromic acid. After cleaning, cuvettes must be cleaned properly with ultrapure water and ethanol and subsequently dried by using a pump dryer or by evaporation.
2. Secondary structure of proteins has signature peaks [28]:

   α-helix: shows negative bands at 216–218 nm and positive band at 195 nm.

   β-sheet: shows negative band at 225 nm and strong positive band between 200 and 205 nm.

   Random coil: shows positive band at 218 nm negative band less than 200 nm.

3. The high tension (HT) voltage must not go beyond 600 V. Otherwise CD signal will be very noisy and the spectral measurements are non-reliable [29, 30].
4. It can analyze only optically active molecules.
5. The concentration of the sample should be known before the start of the CD experiment.

References

1. Strickland EH (1974) Aromatic contributions to circular dichroism spectra of proteins. CRC Crit Rev Biochem 2:113–175
2. Woody RW (1995) Circular dichroism. Methods Enzymol 246:34–71
3. Yang JT, Wu CS, Martinez HM (1986) Calculation of protein conformation from circular dichroism. Methods Enzymol 130:208–269
4. Johnson WC (1988) Proteins through circular dichroism. Ann Rev Biophys Biophys Cher 17: 145–166
5. Thünemann AF, Beyermann J, von Ferber C et al (2000) Immobilization of retinoic acid by polyamino acids: lamellar-structured nanoparticles. Langmuir 16:850–857
6. Johnson WCJ (1985) Circular dichroism and its empirical application to biopolymers. Methods Biochem Anal 31:61–163
7. Johnson WCJ (1990) Protein secondary structure and circular dichroism: a practical guide. Proteins 7:205–214
8. Saikia J, Bhat VT, Potnuru LR et al (2022) Minimalist de novo design of an artificial enzyme. ACS Omega 7:19131–19140
9. Goyal R, Jerath G, Chandrasekharan A et al (2022) Molecular hybridization combining tumor homing and penetrating peptide domains for cellular targeting. Drug Deliv Transl Res 12:1285–1292
10. Prakash V, Mukesh B, Sasidharan S et al (2022) Syndiotactic hexamer peptide nanodots. Eur Biophys J 51:483–491
11. Saikia J, Pandey G, Sasidharan S et al (2019) Electric field disruption of amyloid aggregation: potential noninvasive therapy for Alzheimer's disease. ACS Chem Neurosci 10:2250–2262
12. Pandey G, Das PP, Ramakrishnan V (2020) Directive effect of chain length in modulating peptide nano-assemblies. Protein Pept Lett 27: 923–929
13. Jerath G, Goyal R, Trivedi V et al (2019) Syndiotactic peptides for targeted delivery. Acta Biomater 87:130–139
14. Clippingdale AB, Wade JD, Barrow CJ (2001) The amyloid-β peptide and its role in Alzheimer's disease. J Pept Sci 7:227–249
15. Hazam PK, Akhil R, Jerath G et al (2019) Topological effects on the designability and bactericidal potency of antimicrobial peptides. Biophys Chem 248:1–8
16. Goyal R, Jerath G, Chandrasekharan A et al (2020) Peptide-based delivery vectors with pre-defined geometrical locks. RSC Med Chem 11:1303–1313
17. Prakash V, Ranbhor R, Ramakrishnan V (2020) De novo designed Heterochiral blue fluorescent protein. ACS Omega 5:26382–26388
18. Goyal R, Jerath G, Akhil R et al (2021) Geometry encoded functional programming of tumor homing peptides for targeted drug delivery. J Control Release 333:16–27
19. Saikia J, Dharmalingam K, Anandalakshmi R et al (2021) Electric field modulated peptide based hydrogel nanocatalysts. Soft Matter 17: 9725–9735
20. Gill SC, von Hippel PH (1989) Calculation of protein extinction coefficients from amino acid sequence data. Anal Biochem 182:319–326
21. Pace CN, Vajdos F, Fee L et al (1995) How to measure and predict the molar absorption coefficient of a protein. Protein Sci 4:2411–2423
22. Martin SR, Bayley PM (2002) Absorption and circular dichroism spectroscopy. Methods Mol Biol 173:43–55
23. Hennessey JPJ, Johnson WCJ (1982) Experimental errors and their effect on analyzing circular dichroism spectra of proteins. Anal Biochem 125:177–188
24. Greenfield NJ (1996) Methods to estimate the conformation of proteins and polypeptides from circular dichroism data. Anal Biochem 235:1–10
25. Sreerama N, Woody RW (1994) Protein secondary structure from circular dichroism spectroscopy. Combining variable selection principle and cluster analysis with neural

network, ridge regression and self-consistent methods. J Mol Biol 242:497–507
26. Manning MC (1989) Underlying assumptions in the estimation of secondary structure content in proteins by circular dichroism spectroscopy--a critical review. J Pharm Biomed Anal 7: 1103–1119
27. Greenfield NJ (2006) Using circular dichroism spectra to estimate protein secondary structure. Nat Protoc 1:2876–2890
28. Wallace BA (2009) Protein characterisation by synchrotron radiation circular dichroism spectroscopy. Q Rev Biophys 42:317–370
29. Siligardi G, Hughes CS, Hussain R (2018) Characterisation of sensor kinase by CD spectroscopy: golden rules and tips. Biochem Soc Trans 46:1627–1642
30. Wang X, Yan Y (2023) Optical activity of solids from first principles. Phys Rev B 107:45201

Chapter 7

Fourier Transform Infrared (FTIR) Spectroscopy

Kalpana Kumari and Vibin Ramakrishnan

Abstract

Fourier-transform infrared (FTIR) spectroscopy has been extensively used in chemistry, chemical engineering, and biotechnology. Protein secondary structures can be broadly classified as α-helices, β-sheets, turns, and random coils. The patterns of hydrogen bonding between amide hydrogen and carbonyl oxygen atoms constituting the peptide backbone of a protein are different in each case. The positions of CO and NH stretching and bending vibrations are sensitive to the secondary structure type. In protein structural analysis, it can be used as a useful predictor for secondary structure characterization.

Key words Functional groups, Frequency, Secondary structure, Amide band, Absorption band

1 Introduction

Protein conformation and stability critically depends on its secondary structure constitution and overall fold. Protein infrared (IR) spectrum can possibly have nine amide bands from its main chain and sidechains [1–4]. The most widely characterized bands to verify protein secondary structure are amide I and amide II bands. The absorptions corresponding to C=O stretching are denoted as amide I and N–H bending results in amide II bands. Both C=O and N–H bands are involved in the characteristic repetitive hydrogen bonding interactions between constituent groups in protein main chain. The strength and nature of hydrogen bonding can therefore influence the C=O stretching and N-H vibrations [5–9]. Positions of both amide I and amide II bands are different in different secondary structures, especially between alpha helices and beta sheets. However, IR is widely viewed only as a supporting tool for quantifying the secondary structure of the protein. Yet, FTIR has several advantages such as better speed and sensitivity, high signal-to-noise (S/N) ratio, and relatively simple instrument design. In this chapter, we discuss how this technique has been used to determine the detailed secondary structure of peptides.

Vibin Ramakrishnan (ed.), *Biophysical Characterization of Functional Peptides*, Springer Protocols Handbooks,
https://doi.org/10.1007/978-1-0716-3405-9_7,

The absorption band of IR is in the range of 4000 cm^{-1} to 1500 cm^{-1} to study the functional group. Hence, this region is called as functional group region of IR spectrum [10]. Amide I band is routinely used to determine the conformation of polypeptide backbone [3, 11]. The vibrational frequencies characteristic of protein secondary structures elements are the following [12, 13]:

(a) α-helix: 1657 cm^{-1} to 1648 cm^{-1},
(b) β-sheet: 1641 $cm^{-1}$to 1623 cm^{-1},
(c) Unordered: 1657 $cm^{-1}$to 1642 cm^{-1},
(d) Antiparallel β-sheet: 1695 $cm^{-1}$to 1675 cm^{-1}.

2 Materials

1. A FTIR instrument: Principal suppliers are Perkin- Elmer (Waltham, MA), Jasco (Easton, MD), Bruker Optics (Billerica, MA), and Varian (Palo Alto, CA).
2. 70% Isopropanol.
3. Lint-free tissue.

3 Methods

3.1 Instrument Preparation

1. Instrument should be turned on at least half an hour before collecting the spectra so that it can stabilize. Instrument and sample cell must be equilibrated for 20 min [14].
2. Purge the instrument with N_2 or dry air at a pressure in the range of 20–40 psi. Cool the detector by filling detector cooling system with liq. N_2.
3. Clean the attenuated total reflectance (ATR) crystal gently with 70% isopropanol (*See* **Note 4**).

3.2 Sample Preparation

1. Prepare FTIR sample based on its sampling technique during its spectral acquisition. Peptide must be dissolved in water/deuterium and loaded into the liq. transmission/ATR cells for transmission spectroscopy. Prepare sample as KBr pellets or keep in DRIFTS cell if it is in dry form (*See* **Note 1** and **Note 3**) [15].
2. Thin film of liquid sample is uniformly spread to keep the path length constant (fixed) (*See* **Note 2**).

3.3 Important Points to Note During Data Collection

1. The parameters that need to be adjusted before scanning the samples are the following:
   (a) Number of scans: 100 scans are preferred for high signal-to-noise ratio.

Table 1
Vibrational frequencies of different functional groups [12, 17]

Bond	Molecule	Wavenumber (cm^{-1})
C-O	Alcohol, ether, ester, carboxylic acid	1300–1000
C=O	Aldehyde, ketone, ester, carboxylic acid	1750–1680
C=O	Amide	1680–1630
N-H (stretching)	Amine and amide	3500–3100
N-H (bending)	Amine and amide	1640–1550
O-H	Alcohol	3650–3200
C-N	Amine	1350–1000
S-H	Mercaptan	2550

(b) Spectral range: 4000–600 cm^{-1} is the usually chosen range for most peptides [14].

2. Peltier or external water circulation system is used to control the temperature of FTIR cells.
3. All the parameters are verified by taking few scans without the sample in the sample chamber. Such scans can be used as a background spectrum for the sample. This scan can also be used to verify whether ATR crystal is cleaned or not.
4. Ideally, the buffer should be scanned before sample analysis. It must be subtracted from the scan of sample spectrum.
5. Optimal intensity and high resolution are required to get a high quality FTIR-spectrum.

3.4 Data Analysis

1. Peptide spectrum usually shows multiple bands due to vibrations of peptide bonds. Two most common bands are amide I and amide II peaks [16]. Different regions corresponding to various functional groups are shown in Table 1.

4 Notes

1. KBr must be grinded thoroughly. Otherwise, it may form cloudy pellet due to absorption of moisture that may cause high background noise. So, sample should grind before adding KBr so that there will be less exposure of KBr.
2. There should be just enough solid powder sample in ATR to cover the prism.

3. Avoid water vapor while preparing sample and acquiring the data; otherwise, sharp lines may be present in the sample spectrum.
4. Always use lint-free tissue to clean ATR crystal.

References

1. Goyal R, Jerath G, Akhil R et al (2021) Geometry encoded functional programming of tumor homing peptides for targeted drug delivery. J Control Release 333:16–27
2. Prakash V, Ranbhor R, Ramakrishnan V (2020) De novo designed heterochiral blue fluorescent protein. ACS Omega 5:26382–26388
3. Pandey G, Morla S, Kumar S et al (2020) Modulation of tau protein aggregation using 'Trojan' sequences. Biochim Biophys Acta 1864:129569
4. Pandey G, Morla S, Kumar S et al (2020) Modulating Aβ fibrillogenesis with 'Trojan' peptides. Neuropeptides 81:102030
5. Goyal R, Jerath G, Chandrasekharan A et al (2020) Peptide-based delivery vectors with pre-defined geometrical locks. RSC Med Chem 11:1303–1313
6. Prakash V, Mukesh B, Sasidharan S et al (2022) Syndiotactic hexamer peptide nanodots. Eur Biophys J 51:1–9
7. Saikia J, Dharmalingam K, Anandalakshmi R et al (2021) Electric field modulated peptide based hydrogel nanocatalysts. Soft Matter 17: 9725–9735
8. Saikia J, Pandey G, Sasidharan S et al (2019) Electric field disruption of amyloid aggregation: potential noninvasive therapy for Alzheimer's disease. ACS Chem Nerosci 10:2250–2262
9. Prakash V, Kumari K, Ramakrishnan V (2022) Efficient biosorption of hexavalent chromium from water with human hair, vol 8. ACS Omega, p 915
10. Haris PI, Chapman D (1992) Does Fourier-transform infrared spectroscopy provide useful information on protein structures? Trends Biochem Sci 17:328–333
11. Prakash V, Christian Y, Redkar AS et al (2022) Antibacterial hydrogels of aromatic tripeptides. Soft Matter 18:6360–6371
12. Surewicz WK, Mantsch HH, Chapman D (1993) Determination of protein secondary structure by Fourier transform infrared spectroscopy: a critical assessment. Biochemistry 32:389–394
13. Haris PI, Lee DC, Chapman D (1986) A Fourier transform infrared investigation of the structural differences between ribonuclease a and ribonuclease S. Biochim Biophys Acta Protein Struct Mol Enzymol 874:255–265
14. Bakshi K, Liyanage MR, Volkin DB et al (2014) Fourier transform infrared spectroscopy of peptides. In: Therapeutic peptides. Springer, Berlin, pp 255–269
15. van de Weert M, Haris PI, Hennink WE et al (2001) Fourier transform infrared spectrometric analysis of protein conformation: effect of sampling method and stress factors. Anal Biochem 297:160–169
16. Vass E, Hollósi M, Besson F et al (2003) Vibrational spectroscopic detection of beta-and gamma-turns in synthetic and natural peptides and proteins. Chem Rev 103:1917–1954
17. Ganim Z, Chung HS, Smith AW et al (2008) Amide I two-dimensional infrared spectroscopy of proteins. Acc Chem Res 41:432–441

Chapter 8

Cell Culture for Cell Penetrating Peptides

Yvonne Christian, Dikshita Hazarika, and Vibin Ramakrishnan

Abstract

Animal cell culture is a vital part in basic as well as application oriented research initiatives in biological sciences. It is basically referred as a technique where cells from a tissue are isolated and cultured under controlled conditions in vitro. The homogeneity of cell lines along with the reliability in the generated data makes this technique very useful for making important conclusions. Handling cell culture requires training in employing systematic procedures and maintaining an aseptic environment. This chapter discusses the basics of human cell culture along with handling techniques.

Key words Cell culture, Sterile, Media, In vitro, Passage, Cell line

1 Introduction

Animal cells can be made to grow outside the organ or body when they are provided proper nutrients and physiological like environment. Cell culture as a technique was discovered in the early twentieth century, but later with the development of defined media and transformed cell lines, it became an indispensable tool in biological research [1]. Cell culture technique can include animal, human, or insect cells to be grown and cultured in the lab. The source of the cell line for cell culture can be either a normal tissue like a skin biopsy, or a disease tissue like pancreatic tumor biopsy. Cells isolated from a tissue are called a primary culture which after the first passage is called a cell line. Primary culture have a finite lifespan and undergo senescence very soon. Primary culture acts as a good representative of the cells in vivo, in comparison to the transformed cells [2]. Continuous cell lines have the ability to divide indefinitely as they are transformed by viral oncogenes such as SV40T or by telomerase induction. Continuous cell lines are homogenous in nature; provide a continuous supply of cells resulting in good levels of consistency in experimental results [3]. Cell lines can be of two types: adherent and suspension. Adherent cell lines are

Vibin Ramakrishnan (ed.), *Biophysical Characterization of Functional Peptides*, Springer Protocols Handbooks,
https://doi.org/10.1007/978-1-0716-3405-9_8,

anchorage-dependent and require a solid surface for the attachment and growth. Once the surface is covered completely by a monolayer of cells they stop dividing. An example is the human embryonic kidney cell line HEK-293 which is an epithelial cell, whereas suspension cells do not require a surface for attachment and grow freely in the liquid media. THP-1, a monocytic cell line is an example of a suspension cell line.

Three basic procedures are followed while performing cell culture. The first one is the revival and initiation of the cell line. Cell lines appropriate for a certain type of research can be obtained from cell banks. These are thawed at 37 °C and placed in a container such as a flask or petri dish containing nutrient media. Once the cells have attached and divided and covered 80% of the surface of the flask or dish they are said to be 80% confluent. At this point they are detached from the surface using the enzyme trypsin and diluted into a second flask or dish. This process is known as passaging or sub-culture. Lastly in order to preserve stocks of the cell line in a stable state, a freezing mixture with a cryoprotectant like dimethyl sulfoxide (DMSO) is used. Cells in the freezing mixture can be stored for many days in liquid nitrogen which has a temperature of −196 °C [4, 5].

Maintaining an aseptic environment is the defining criteria for a cell culture laboratory. If cells are not cultured in a sterile aseptic condition then contamination results and the culture needs to be discarded. For this reason, a laminar hood is used which maintains an aseptic environment for work. Sterile media, dissociation reagents, and phosphate buffer saline are commercially available for use. These must only be opened in a laminar hood and not outside. Flasks, pipettes, petri dishes, and multi-well plates which are sterilized by gamma radiation are used. The surfaces of the materials are wiped with a solution of 70% ethanol. Above all, the person working with cell culture must maintain personal good hygiene. Washing of hands with soap and water before and after cell culture is an important basic habit to be followed religiously.

There are a lot of advantages of using in vitro cell lines in biological research. Cell cultures are used to study factors affecting growth, cell division, metabolism, genetics, pharmacology, and toxicology. As a result of homogeneity in the cells, the biggest advantage of using cell culture is the reproducibility of data. Diseased cancer cell lines like MDA-MB-231 (for breast cancer), HeLa (for cervical cancer), and A549 (for non-small cell lung carcinoma) help in understanding cancer cells as well testing of various anti-cancer molecules. The cellular uptake and drug delivery potential of syndiotactic cell-penetrating peptides attached to an anti-cancer drug methotrexate has been studied in cancerous (MDA-MB-231 and HeLa) and non-cancerous cell lines (HEK-293) [6, 7]. Novel tumor homing peptides capable of specifically homing to receptors on cancer cells have been designed, synthesized, and studied in cancerous (MDA-MB-231, HeLa, and U2-OS) and non-cancerous cell lines (MCF-10A). Their drug delivery potential was compared in in vivo mouse models and the results obtained were identical and comparable to in vitro cell culture studies [8]. Moreover, a drug delivery unit comprising of cell-penetrating peptide with a tumor homing domain was assessed for uptake and cytotoxicity in cell lines MDA-MB-231, HeLa, and HEK-293 [9]. Commercially available biological products like vaccine, monoclonal antibodies, and functional proteins with post-translational modifications are produced using animal cell culture. Gene therapy and toxicity effects of new drugs are also first studied on cell lines [2, 10].

There are certain limitations associated with this technique. In cell culture, only a monolayer of cells is being used to study the biological processes or the effect of a drug molecule. So, it cannot compensate for studies that are performed in vivo in animals where different physiological phenomena are taken into consideration. Cell culture is a specialized technique and requires skilled persons and costly equipment which makes it an expensive technique. Genotypic and phenotypic instability arises due to increase in the passage number of the cells leading to properties differing from the original cells. Contamination is a major problem in cell culture which arises due to organisms like yeast, bacteria, and mold if strict aseptic conditions are not followed. Such contaminants can be detected by observation under an inverted microscope, turbidity of media, and change in the pH of the media indicated by a change in the color of phenol red indicator from red to yellow. These contaminants can be prevented by use of antibiotics within a certain limit. Using higher doses of antibiotic can give rise to antibiotic-resistant strains. One of the more problematic contaminations is due to mycoplasma which is not detected through the microscope and does not show any morphological change in the cells and therefore it usually goes undetected. They can alter cellular metabolism and growth leading to erroneous experimental results. Mycoplasma are unaffected by the regular antibiotics used in cell culture like penicillin and many species are smaller than 0.2 μm and cannot be filtered. Detection of mycoplasma contamination is possible through PCR and ELISA based assays or through selective microbial agar and broth [11]. Cross-contamination of a different cell line with the cells of our culture is also a problem which can be prevented by procurement of cell lines from reputable cell banks and proper handling techniques [12].

1.1 Biosafety

Occupationally acquired infection and release of pathogens into the surrounding needs to be prevented during the cell culture technique and this can be achieved only through a thorough understanding of the concept of biological safety. Hazardous biological materials include human pathogens like virus, bacteria, fungi and prions, human blood and body-fluids, tissues and cultured cells [12].

Varying from low risk to high risk, biosafety can be classified into four different levels [12]:

(a) *Biosafety Level 1*: Associated with low risk where organisms that do not cause any human disease are grown.

(b) *Biosafety Level 2*: Here moderate-risk organisms that can cause diseases in humans through ingestion or percutaneous membrane contact are cultured. This requires working in Biosafety Class II cabinets.

(c) *Biosafety Level 3:* This level is associated with moderate to high-risk organisms capable of respiratory transmission and causing lethal infections. This requires working in Biosafety Class III cabinets.

(d) *Biosafety Level 4:* Associated with highest risk levels where organisms transmitted through aerosol route and capable of causing life threatening diseases with no cure or vaccine available are cultured. Personnel working in such labs have to wear a full body positive pressure protective suit.

The waste generated by the culture of cells and organisms must be discarded in proper way so as to prevent contaminants escaping in the surrounding environment. Tissue culture medium waste must be inactivated with a solution of sodium hypochlorite prior to disposal and other materials such as flasks, pipettes, and petri dishes must be disposed in specific biohazard bags and incinerated.

1.2 Tissue Culture Laboratory: Design and Layout

Sterility is the major factor which makes a tissue culture laboratory different from any other common laboratory space. The culture room must be separate from other working areas with restricted entry of individuals. It should be cleaned regularly to minimize dust particles. The lab can also be fitted with an ultraviolet (UV) light for complete sterilization of the area when not in use. Also, it must be fumigated from time to time using a mixture of potassium permanganate and formaldehyde (*see* **Note 1**). Preparation, handling, incubation, washing, sterilizing, and storage are the major steps that must be performed in a tissue culture laboratory.

Basic equipment required for works in a cell culture lab are:

(a) Laminar Air Flow: A laminar air flow hood is used to eliminate the risk of contaminants in our culture. It contains a highly effective filter known as high efficiency particulate air (HEPA) filter which filters out dust particles and bacteria and controlled air flow that prevents contamination in the culture. Laminar hoods are of two kinds, namely vertical and horizontal. Both types of hoods maintain sterility of the work area. But in a horizontal hood the flow of air is directed towards the

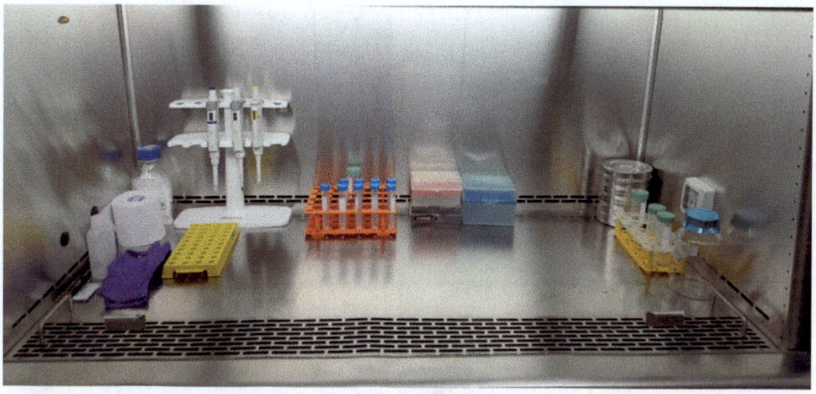

Fig. 1 Arrangement of materials inside a laminar flow hood

user, whereas in a vertical hood the air flow is such that it protects the working personnel from coming in contact with the samples inside. Therefore, for cell culture, horizontal laminar hoods are not used. Before starting cell culture work, the area inside the hood is wiped with 70% ethanol. The hood is fitted with a UV light inside which is switched on for a minimum of 20 min prior to work maintaining additional sterility [4, 5] (*see* **Note 2**) (Fig. 1).

(b) Incubator: A CO_2 incubator maintains three parameters temperature, CO_2 and relative humidity to replicate the in vivo environment in the culture. All the cells and tissues seeded in flasks and petri dishes are maintained in various racks inside the incubator. Temperature is maintained at the physiological temperature of 37 °C. CO_2 level is regulated at 5% to maintain the pH of the culture medium. 95% humidity is maintained by a pan filled with distilled water containing a small amount of copper sulphate ($CuSO_4$) which acts as an antibacterial and antifungal agent (Fig. 2).

(c) Microscope: A light microscope is used to monitor the growth and morphology of the cells in culture. The microscope is an inverted one where the objective is below the stage to allow visualization of cells in a flask, petri dish, or multi-well plate. The microscope is also used to check for contamination in the culture. Usually 10×, 20×, and 40× objectives are sufficient. Moreover, it is also beneficial to have phase-contrast optics and a camera attached to the microscope [11] (Fig. 3).

(d) Centrifuge: A table-top centrifuge is required to pellet out the cells from the media after trypsinization. It contains slots for centrifugation of tubes and smaller vials with both RPM (revolutions per minute) and RCF (relative centrifugal force) modes (Fig. 4).

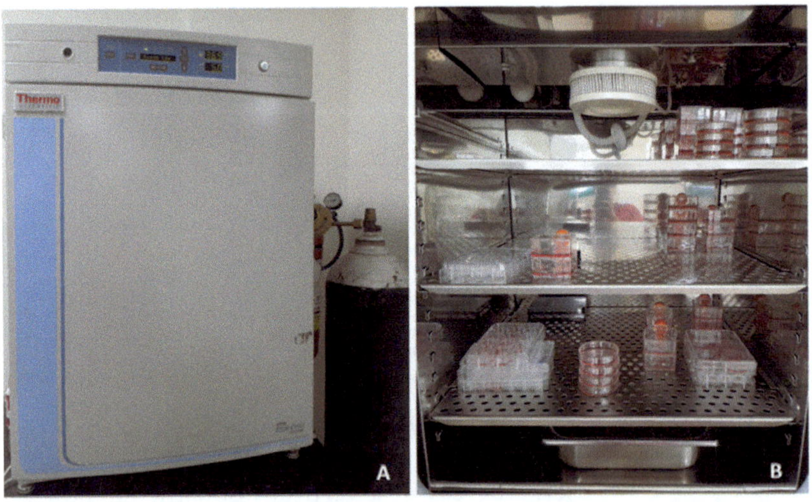

Fig. 2 (**a**) CO_2 incubator attached to a CO_2 cylinder. (**b**) Flasks, petri dishes, and multi-well plates containing cell cultures placed on racks inside the incubator

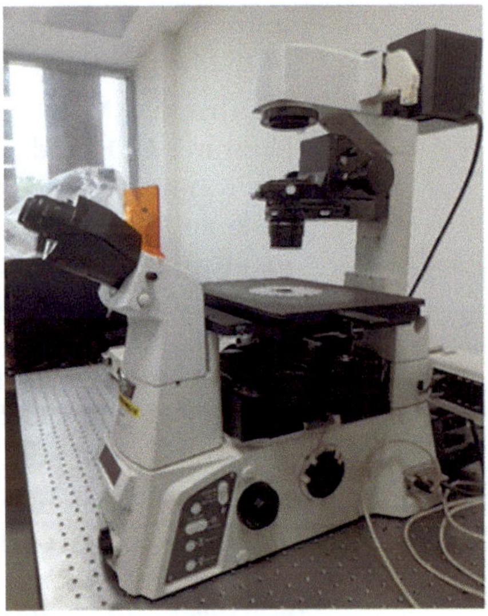

Fig. 3 Inverted microscope

(e) Refrigerator*:* In a cell culture lab, a refrigerator with temperature of 4 °C is used to store culture media and a freezer with −20 °C is used to store antibiotic and serum bottles (Fig. 5).

(f) Water Bath: Media bottles stored in the refrigerator at 4 °C are kept in a water bath set at 37 °C prior to use in cell culture to prevent giving a temperature shock to the cells (Fig. 6).

All the above equipment must be kept clean with maintenance checks performed regularly.

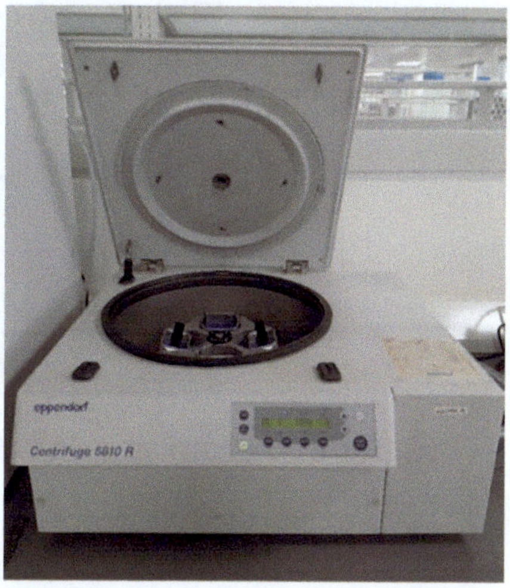

Fig. 4 Table top centrifuge

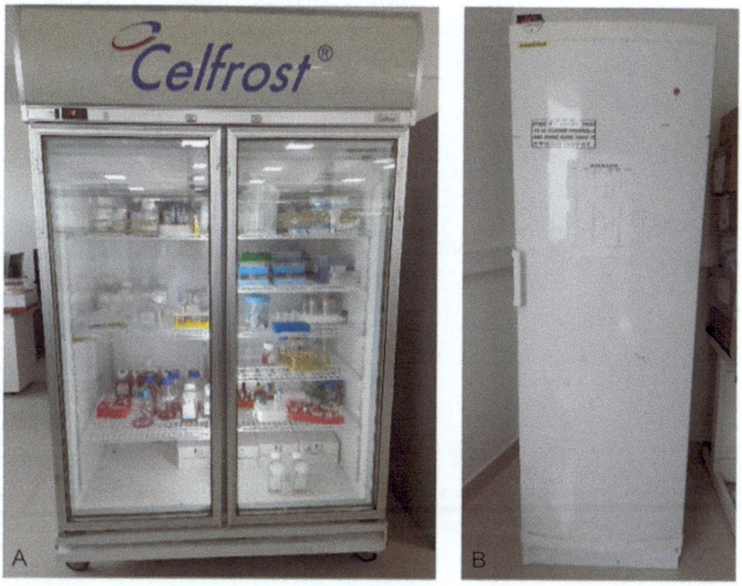

Fig. 5 (**a**) Refrigerator set at 4 °C to store media, PBS. (**b**) Freezer set at −20 °C for storing antibiotic, trypsin, and other reagents

2 Materials

(a) Media: Depending on the type of cells to culture the type of media varies. Three varieties of basic media are available, namely Eagle's minimal essential medium (EMEM), Dulbecco's modified Eagle's medium (DMEM), and Roswell Park Memorial Institute (RPMI) medium 1640. These contain

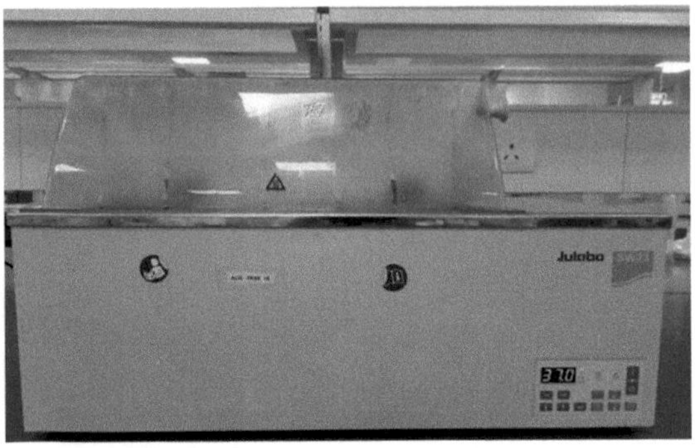

Fig. 6 Water bath set at 37 °C for thawing of media and cryopreserved culture vials

various amino acids, vitamins, salts, glucose, and other nutrients. Media formulations are available from commercial suppliers in a powder or liquid form. The liquid form is in a ready to use sterile form and is very convenient. But it is expensive. The powdered form of media has to be dissolved in sterile water and then filtered sterilized using a 0.22 μm filter inside a laminar hood. A basic medium is generally not enough for the growth of cells and is usually supplemented with a variety of components. Serum is a major supplement used in the media which contains growth factors for cell proliferation and maintenance. Mostly fetal bovine serum (FBS) is used. It is used at a concentration of 10% in the basic medium. The pH of the media must be in 7.2–7.4 range, which is maintained by addition of buffers like sodium bicarbonate ($NaHCO_3$) and 4-(2-hydroxyethyl)-1-piperazineethanesulfonic acid (HEPES). Phenol red is added as a pH indicator. Moreover, the amino acid L-glutamine is added which is essential for the synthesis of proteins and nucleic acids. Antibiotics like penicillin and streptomycin are also added in the media to prevent the growth of contamination. Basic media along with the components mentioned above make up the "complete media" [5, 12].

(b) Phosphate Buffer Saline (PBS): This buffer with a pH of 7.4 is used in the rinsing of cells between media change.

(c) Trypsin-EDTA Solution: Trypsin is a protease enzyme used to detach cells and EDTA acts as a chelator. Usually a concentration of 0.05–0.5% trypsin solution is used in culture. The time required for detachment by trypsin is crucial as treatment for a longer time can cause damage to the cells. Temperature of 37 °C is required for the enzyme activity.

(d) 70% Ethanol: A solution of 70% ethanol is prepared in sterile water and is used to wipe the surface of the laminar hood as well as all the supplies that are placed inside the hood.

(e) Flasks (T-25, T-75), Petri Dishes (100 mm, 60 mm), Multi-well Plates (6-well, 12-well, 24-well, 48-well, 96-well): These are transparent, sterile, made of polystyrene, and treated with poly-D-lysine or collagen for cell attachment.

(f) Sterile pipettes.

(g) Sterile centrifuge tubes.

(h) Neubauer's Chamber: For counting cells.

(i) Autoclave: For sterilizing bottles and pipette tips.

(j) Cryovials.

(k) DMSO: DMSO acts as a cryoprotectant as it reduces the freezing point of the media and reduces formation of ice crystals which can damage the cell. A solution of 10% DMSO in the media is used for cryopreservation of cells [5].

(l) Cryo containers.

(m) Isopropanol.

3 Methods

3.1 Cell Revival

(a) Switch on the laminar hood and wipe the surface and working area of the hood with 70% ethanol. Switch on the UV light for 20 min.

(b) Pre-warm the media bottles in a 37 °C water bath. Wipe the outer surface of the bottle with 70% ethanol and place it in the laminar hood after UV sterilization is over.

(c) Aliquot 5 mL of media into a T-25 flask.

(d) Remove the cryopreserved frozen cell line from the liquid nitrogen tank (*see* **Note 3**).

(e) Immediately place it for thawing in a water bath set at 37 °C.

(f) Remove the vial from the water bath once the mixture has been 80% thawed.

(g) Wipe the surface of the vial thoroughly with 70% ethanol and place it in the laminar hood.

(h) With the help of a pipette aliquot out the thawed cells and gently pipette it into the T-25 flask containing the media. Pour out the mixture very slowly drop by drop (*see* **Note 4**).

(i) Incubate the flask at 37 °C in a CO_2 incubator overnight.

(j) In the next morning, check to see if the cells have attached to the surface of the flask by the help of an inverted microscope.

(k) Discard the media in the flask and add 5 mL of fresh media. Incubate at 37 °C in a CO_2 incubator.

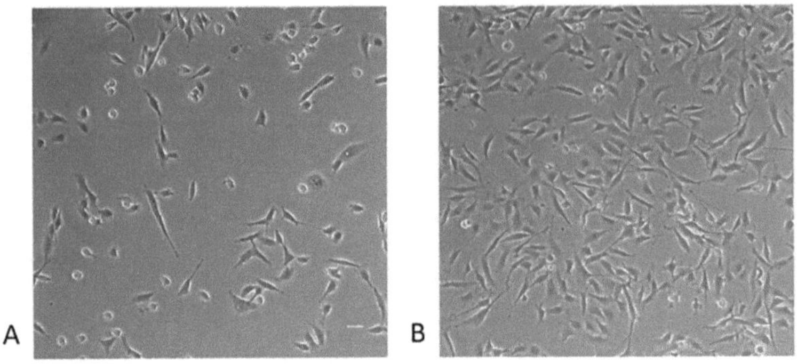

Fig. 7 MDA-MB-231 cells observed under an inverted phase-contrast microscope. (**a**) Cells are less in number and less confluent. (**b**) Cells are approximately 85% confluent and ready for sub-culture

(l) Once the confluency of the cells reaches around 80%, proceed for sub-culturing.

3.2 Sub-Culturing of Adherent Cells

(a) Once cells reach a confluency of around 80%, proceed for sub-culture (Fig. 7).

(b) In the laminar hood, remove the media from the flask.

(c) Wash the flask with 3 mL PBS (*see* **Note 5**).

(d) Add 1–2 mL of trypsin-EDTA solution to cover the entire surface of the flask.

(e) Close the flask and place it in the CO_2 incubator at 37 °C till the cells are detached (*see* **Note 6**).

(f) Once the cells are seen to be detached, immediately add 2 mL of complete media. The serum in the media deactivates the enzymatic activity of trypsin.

(g) Transfer the components in the flask to a sterile centrifuge tube.

(h) Centrifuge the cell solution at 1000 rpm for 5 min at room temperature.

(i) Discard the supernatant and wash the pellet with PBS and proceed to centrifuge as mentioned in the previous step (*see* **Note 7**).

(j) Discard the supernatant and resuspend the pellet in complete media.

(k) Count the cell number using Neubauer's chamber (*see* **Note 8**).

(l) Add media into a new flask and add half of the cell suspension to get 1:2 sub-culture ratio. A ratio of 1:3 or 1:4 can also be prepared.

(m) Note down the passage number, cell line, and date on the flask and place it in the CO_2 incubator.

(n) The cultured dish must be then incubated in a CO_2 incubator.

(o) Discard the old media and used pipettes and tubes appropriately. Place the media, trypsin, and PBS bottles in the refrigerator. Wipe the surface of the laminar hood using 70% ethanol.

In case of suspension cell culture, the procedure of sub-culturing becomes easier as there is no need for trypsinization.

3.3 Media Change

(a) Pre-warm the media bottles in a water bath set at 37 °C.

(b) Wipe the surface of the media bottle with 70% ethanol before placing in the hood.

(c) Discard the media from the flask using a pipette.

(d) Wash the flask with 3 mL PBS buffer by gentle swirling. Discard the PBS.

(e) Add 5 mL of the media to the flask. Cap the flask and place it in the CO_2 incubator at 37 °C.

(f) Discard the media and PBS waste and the used pipettes appropriately.

(g) Place the media and PBS bottles in the fridge for future use.

(h) Swipe the laminar hood surface with 70% ethanol.

3.4 Cryopreservation of Cells

(a) Check the growth and condition of cells growing in a flask under the microscope. Cells must be in the log phase and contamination must be absent.

(b) Discard the media from the flask, wash once with PBS, and trypsinize the cells. Centrifuge to obtain the cell pellet.

(c) Count the cells using Neubauer's chamber.

(d) Prepare the freezing mixture containing complete media with 10% DMSO.

(e) Re-suspend 1×10^6 cells in 1 mL of the freezing mixture and transfer it into sterile cryovials.

(f) Cap the vials immediately and label them with the name of the cell line and the date.

(g) Quickly place the vials into cryo container which aids in cooling the vials at the rate of 1 °C per minute (*see* **Note 9**).

(h) Transfer the box into −80 °C freezer and keep it overnight.

(i) Transfer the vials into a tank containing liquid nitrogen the next day.

4 Notes

1. The fumes coming out during fumigation are extremely toxic. Care must be taken not to inhale them and stay outside the room.
2. Laminar hood UV is switched on for a minimum 20 min prior to the cell culture work. After finishing work, it is advised to switch on the UV for 20 min before switching off the hood.
3. Wear protective insulating gloves while removing vials from liquid nitrogen.
4. Gentle pipetting must be practiced because the cryopreserved cells are in delicate and stressed condition.
5. Rinsing with PBS removes dead cells and the toxins produced due to cell metabolism.
6. Keeping the cells in the trypsin dissociation reagent for an extended time may be avoided. The dissociation by trypsin for each cell line is different based on its properties.
7. Centrifugation step after trypsinization ensures the removal of traces of the enzyme.
8. Counting cells in a Neubauer's chamber is necessary when a certain amount of cells need to be seeded in a well. Cells stained with trypan blue are considered dead cells and the unstained cells are viable cells. Cells are counted from the four corner squares in the chamber. Cells per mL are determined by the equation:

$$\text{cells/mL} = \text{average count per square} \times \text{dilution factor} \times 10^4 \quad (1)$$

9. The cryo containers are filled with isopropanol which aids in the controlled rate of cooling.

References

1. Freshney RI (1994) Culture of animal cells: a manual of basic technique, 3rd edn. Wiley-Liss, New York
2. Verma A, Verma M, Singh A (2020) Animal tissue culture principles and applications. Anim Biotechnol:269–293
3. Kaur G, Dufour JM (2012) Cell lines: valuable tools or useless artifacts. Spermatogenesis 2(1): 1–5
4. Coecke S, Balls M, Bowe G, Davis J, Gstraunthaler G, Hartung T, Hay R, Merten OW, Price A, Schechtman L, Stacey G, Stokes W (2005) Second ECVAM task force on good cell culture practice. Guidance on good cell culture practice a report of the second ECVAM task force on good cell culture practice. Altern Lab Anim 33(3):261–287
5. Phelan K, May KM (2017) Mammalian cell tissue culture. Curr Protoc Hum Genet 94: a.3G.1–a.3G.22
6. Jerath G, Darvin P, Christian Y, Trivedi V, Kumar TRS, Ramakrishnan V (2022) Delivery of small molecules by syndiotactic peptides for breast cancer therapy. Mol Pharm 19(8): 2877–2887

7. Jerath G, Goyal R, Trivedi V, Santhoshkumar TR, Ramakrishnan V (2019) Syndiotactic peptides for targeted delivery. Acta Biomater 87: 130–139
8. Goyal R, Jerath G, Akhil R, Chandrasekharan A, Puppala ER, Ponneganti S, Sarma A, Naidu VGM, Santhoshkumar TR, Ramakrishnan V (2021) Geometry encoded functional programming of tumor homing peptides for targeted drug delivery. J Control Release 333:16–27
9. Goyal R, Jerath G, Chandrasekharan A, Christian Y, Kumar TRS, Ramakrishnan V (2022) Molecular hybridization combining tumor homing and penetrating peptide domains for cellular targeting. Drug Deliv Transl Res 12(5):1285–1292
10. Segeritz CP, Vallier L (2017) Cell culture: growing cells as model systems in vitro. In: Basic science methods for clinical researchers, pp 151–172
11. Geraghty RJ, Capes-Davis A, Davis JM, Downward J, Freshney RI, Knezevic I, Lovell-Badge R, Masters JR, Meredith J, Stacey GN, Thraves P, Vias M (2014) Cancer Research UK. Guidelines for the use of cell lines in biomedical research. Br J Cancer 111(6): 1021–1046
12. Philippeos C, Hughes RD, Dhawan A, Mitry RR (2012) Introduction to cell culture. Methods Mol Biol 806:1–13

Chapter 9

Comparative Uptake of Peptides by Flow Cytometry

Yvonne Christian and Vibin Ramakrishnan

Abstract

Cell-penetrating peptides (CPPs) offer great potential as delivery systems of various drugs and imaging agents for a variety of diseases. They have been developed as an alternative to toxic methods like electroporation, drugs encapsulated in polymers, and viral drug delivery. To verify the utility of CPPs as a delivery vehicle, first and foremost, the cellular uptake of the peptides has to be studied. A fluorophore is covalently attached to the CPP molecule, which helps to study the internalization of the CPP molecule in the cell. The uptake can be quantified by the method of flow cytometry which is highly sensitive. We have tagged our de novo designed peptide with the fluorophore 5(6)-carboxyfluorescein and treated various cancerous (MDA-MB-231, HeLa, A549) and non-cancerous cell lines (HEK-293) with a fixed concentration of de novo designed peptide for a fixed time. The cells are then subjected to analysis by flow cytometry and the differential uptake in the cell lines can be quantified.

Key words Cell-penetrating peptides (CPPs), Cellular uptake, Fluorophore, Flow cytometry

1 Introduction

The hydrophilic nature of various therapeutic molecules used for diseases such as cancer limits their uptake through the cellular membrane. Cell-penetrating peptides (CPPs) are short peptides comprising of 5–30 amino acids, capable of traversing through the plasma membrane. In the year 1988, two independent groups discovered that transactivator of transcription (TAT) protein of HIV-1 could be rapidly taken up by cells in vitro [1, 2]. In the year 1991, the peptide pAntp corresponding to the antennapedia homeodomain in Drosophila was found to penetrate into differentiated neuronal cells and accumulate in their nuclei [3]. A shorter fragment of 16 amino acids from the third helix of pAntp protein could also internalize into cells. Consequently, a significant amount of research was done in discovering new CPPs or developing new synthetic CPP molecules. CPPs can be classified into three types based on their origin, namely natural peptides, model peptides, and designed peptides. Natural peptides are the peptides derived from

Vibin Ramakrishnan (ed.), *Biophysical Characterization of Functional Peptides*, Springer Protocols Handbooks, https://doi.org/10.1007/978-1-0716-3405-9_9,

naturally occurring protein-transduction domain (PTD) and they include the TAT and Penetratin peptides. Model peptides have repeating units of amino acids in their sequence and R9 peptide is an example of this class. Designed peptides are novel peptides that are derived from fusion of certain protein or peptide residues or by rational mutations. For example, peptides MPG, MPG-α, and Pep-1 combine a fusion peptide or a tryptophan-rich segment with a nuclear localization sequence [4].

There are two routes of cellular uptake for a CPP, an energy independent route and an energy dependent route. The energy independent route is through direct translocation across the cellular membrane and CPPs that are cationic in nature take this route for internalization. This route is opted when the CPP molecules interact with the membrane, causing either membrane destabilization or transient pore formation. The membrane destabilization is through the "carpet" model which increases membrane fluidity causing the CPPs to enter. The "inverted micelle" model outlines the possibility of CPPs entering through invagination of the cell membrane [5]. The transient pore formation is also described by two models, namely the "barrel-stave pore" model and the "toroidal-pore" model. In the "barrel-stave pore" model the CPP assumes an α-helical conformation while interacting with the membrane and its hydrophilic side chains form the inner part of the transient pore. In the "toroidal-pore" model, the peptides interact with the polar head groups of the membrane, causing a rearrangement and formation of a pore [6]. The energy dependent route by the naturally occurring process in all cells is termed as endocytosis. It is mainly of three types: clathrin/caveolae dependent, clathrin/caveolae independent, and macropinocytosis. In this process the CPPs are trafficked in organelles called endosomes and in order for them to reach the cytosol or a target organelle they need to be released from the endosomes. Endosomal entrapment of CPPs is a major barrier preventing CPPs from reaching the cytosol and performing their function. Several novel approaches have been developed to enhance the endosomal escape of CPPs. Although a lot of research has been done, the exact mechanism of CPP uptake is not yet known and a CPP can enter a cell by more than one route. This depends on the size, nature, and concentration of the CPP as well its cargo [7].

To study the cellular uptake of CPP various methods have been developed. The most common way is to label the peptide molecule with a fluorophore in order to track its entry and localization into the cell. In one study a drug delivery peptide vector containing a syndiotactic CPP was tagged covalently with the fluorophore 5(6)-carboxyfluorescein and its internalization in various cell lines was quantified by fluorescence spectroscopy using a multimode microplate reader [8]. Qualitative internalization of a drug delivery peptide vector containing a syndiotactic CPP and a tumor homing domain liked to a fluorophore can also be studied by treating cells with the peptide-fluorophore conjugate and visualized through confocal microscopy [9]. Organelles like endosomes and lysosomes were stained with commercially available dyes and the colocalization of CPP in these organelles was studied [10]. But to study the uptake of CPP quantitatively, the best and most favorable way is through the method of flow cytometry which detects fluorescence signal with utmost sensitivity [11]. Here fluorescence of intact cells can be measured, which do not require to be lysed. In one study cancerous cells (MDA-MB-231, HeLa, U2-OS) and non-cancerous cells (MCF-10A) were treated with tumor homing peptides and the differential cellular uptake between the cell lines was determined using flow cytometry in the green channel [12]. Another study described the use of short syndiotactic CPPs attached to the drug methotrexate for targeting breast cancer in in vitro and in vivo models. The peptides were tagged with the fluorophore 5(6)-carboxyfluorescein and incubated in cells for 4 h and their uptake was studied by flow cytometry [13]. In yet another example, a series of cationic, amphipathic CPPs displaying a helical conformation were designed for targeting cancerous cells. Cellular uptake of these peptides tagged with a fluorophore was studied by flow cytometry [14]. In one study, the amino acid glycine in the tumor homing peptide motifs, RGD and NGR was replaced with D-proline producing a topologically constrained peptide with enhanced proteolytic stability. The uptake studies of the peptide delivery vector were investigated in cancerous and non-cancerous cell lines in a similar way by flow cytometry [15]. One study involving the investigation of uptake of CPPs TAT and R9 by flow cytometry showed that cells treated with the peptide need to undergo trypsin digestion before FACS analysis. This is done to distinguish between internalized CPP and membrane-bound CPP which is thought to be digested by trypsin [16]. Therefore, flow cytometry presents a very elegant method to quantitatively determine the internalization of peptides.

2 Materials

2.1 Equipment

1. 1300 Series Class II, Type A2 Biological Safety Cabinet (Thermo Fisher Scientific).
2. Galaxy 170 S CO_2 incubator (Eppendorf, New Brunswick).
3. 3K30 centrifuge (Sigma).
4. Fluorescence activated cell sorting (FACS) machine (BD FACS Aria).

2.2 Reagents/Supplies

1. Cell lines: HEK-293, MDA-MB-231, HeLa, and A549 cell lines were purchased from ATCC.
2. Culture media: All the cell lines were cultured in DMEM media (Gibco) supplemented with 10% FBS and 1% penicillin (10,000 units/mL)–streptomycin (10,000 μg/mL).
3. 0.05% Trypsin-EDTA solution for cell dissociation (Gibco).
4. Phosphate-buffered saline (PBS), pH 7.4 (Gibco).
5. 0.4% Trypan blue solution (Gibco).
6. Sterile 24-well plate tissue culture plate with flat bottom (Becton Dickinson).
7. 25 cm^2 rectangular canted neck cell culture flask with vent cap (Corning).
8. 5 mL polystyrene round-bottom tube (Falcon, Corning).
9. CPPs (de novo designed peptide and TAT) were synthesized manually by solid-phase peptide synthesis and attached with the fluorophore 5(6)-carboxyfluorescein at the N-terminal.

3 Method

1. Cell lines HEK-293, MDA-MB-231, HeLa, and A549 are cultured in DMEM media containing 1 mg/mL penicillin-streptomycin antibiotic solution and 10% FBS in T-25 flasks and maintained in an incubator at 37 °C with 5% CO_2 and 95% humidity.
2. For seeding, detach cells from 25 cm^2 flasks using 0.05% Trypsin-EDTA solution and placing in the incubator at 37 °C for 3–5 min. Add media to inactivate trypsinization and centrifuge at 1000 rpm for 5 min at room temperature to obtain cell pellet. Resuspend this pellet in 1 mL of media.
3. Count the number of cells by staining with 0.4% solution of Trypan blue (*see* **Note 1**).
4. Seed the cells in 24-well plates with a seeding density of 70,000 cells per well and incubated for 24 h.
5. The following day, discard the media and wash cells once with PBS.
6. 5 μM peptide solutions are prepared in serum-free media. As a positive control the known cell-penetrating peptide TAT is used (*see* **Note 2**)
7. Treat the cells with peptides having a concentration of 5 μM and incubate the plate in the incubator for 3 h.
8. After incubation, discard the media containing peptide from the wells and wash once with PBS.

9. Add 400 μL of 0.04% Trypan blue to the cells and incubate at room temperature for 5 min (*see* **Note 3**).
10. After 5 min discard the Trypan blue solution and wash the wells thrice with PBS.
11. Add 400 μL trypsin-EDTA solution for 5 min to detach the cells from the surface of the plate. Neutralize by adding 800 μL of complete media.
12. Collect the cells in 5 mL round-bottom tubes and centrifuge at 2000 rpm at 4 °C for 5 min. Discard the supernatant after centrifugation.
13. Resuspend the cells in 500 μL PBS and centrifuge as described above. Discard the supernatant.
14. Resuspend the cells in 200 μL ice-cold PBS solution and place the tubes on ice.
15. Proceed immediately for FACS analysis.
16. The results of individual samples are normalized by subtracting the autofluorescence value from the untreated cells from the sample's fluorescence value. The result of mean fluorescence intensity versus peptides for four different cell lines is displayed as a histogram (Fig. 1).

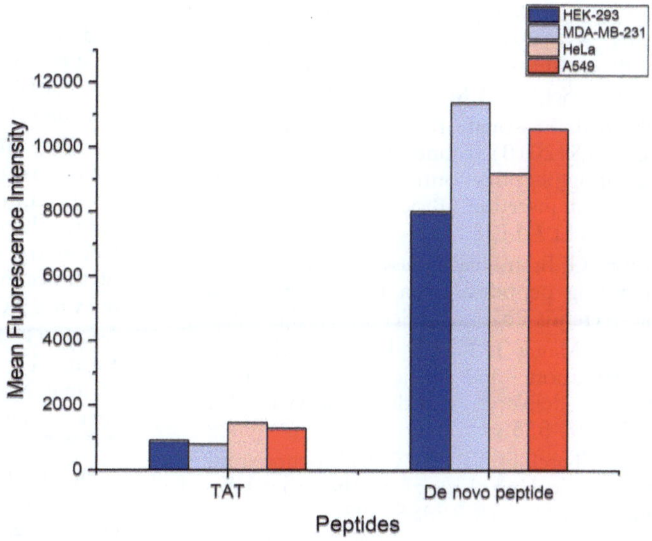

Fig. 1 Quantitative uptake of two peptides, de novo designed and TAT in the cell lines HEK-293, MDA-MB-231, HeLa, and A549. The de novo designed peptide shows higher uptake than TAT

4 Notes

1. To count cells a 1:1 mixture of the cells and Trypan blue solution is prepared and incubated for 5 min. The counting is done with the help of a hemocytometer.
2. Peptides are dissolved in sterile ultrapure water and the solution is passed through 0.2 micron filter in order to maintain sterility. Concentration of peptides is determined by UV-absorption spectrophotometer by taking absorbance at 485 nm for 5,6-carboxyfluorescein and using Beer-Lambert's equation.
3. Trypan blue is a non-permeant stain which is known to quench the green fluorescence of the peptides attached to the outer surface of the cells.

References

1. Frankel AD, Pabo CO (1988) Cellular uptake of the tat protein from human immunodeficiency virus. Cell 55(6):1189–1193
2. Green M, Loewenstein PM (1988) Autonomous functional domains of chemically synthesized human immunodeficiency virus tat trans-activator protein. Cell 55(6):1179–1188
3. Joliot A, Pernelle C, Deagostini-Bazin H, Prochiantz A (1991) Antennapedia homeobox peptide regulates neural morphogenesis. Proc Natl Acad Sci U S A 88(5):1864–1868
4. Eiríksdóttir E, Konate K, Langel U, Divita G, Deshayes S (2010) Secondary structure of cell-penetrating peptides controls membrane interaction and insertion. Biochim Biophys Acta 1798(6):1119–1128
5. Guidotti G, Brambilla L, Rossi D (2017) Cell-penetrating peptides: from basic research to clinics. Trends Pharmacol Sci 38(4):406–424
6. Järver P, Mäger I, Langel Ü (2010) In vivo biodistribution and efficacy of peptide mediated delivery. Trends Pharmacol Sci 31(11):528–535
7. Koren E, Torchilin VP (2012) Cell-penetrating peptides: breaking through to the other side. Trends Mol Med 18(7):385–393
8. Jerath G, Goyal R, Trivedi V, Santhoshkumar TR, Ramakrishnan V (2019) Syndiotactic peptides for targeted delivery. Acta Biomater 87: 130–139
9. Goyal R, Jerath G, Chandrasekharan A, Christian Y, Kumar TRS, Ramakrishnan V (2022) Molecular hybridization combining tumor homing and penetrating peptide domains for cellular targeting. Drug Deliv Transl Res 12(5):1285–1292
10. Gross A, Alborzinia H, Piantavigna S, Martin LL, Wölfl S, Metzler-Nolte N (2015) Vesicular disruption of lysosomal targeting organometallic polyarginine bioconjugates. Metallomics 7(2):371–384
11. Illien F, Rodriguez N, Amoura M, Joliot A, Pallerla M, Cribier S, Burlina F, Sagan S (2016) Quantitative fluorescence spectroscopy and flow cytometry analyses of cell-penetrating peptides internalization pathways: optimization, pitfalls, comparison with mass spectrometry quantification. Sci Rep 6:36938
12. Goyal R, Jerath G, Akhil R, Chandrasekharan A, Puppala ER, Ponneganti S, Sarma A, Naidu VGM, Santhoshkumar TR, Ramakrishnan V (2021) Geometry encoded functional programming of tumor homing peptides for targeted drug delivery. J Control Release 333:16–27
13. Jerath G, Darvin P, Christian Y, Trivedi V, Kumar TRS, Ramakrishnan V (2022) Delivery of small molecules by Syndiotactic peptides for breast cancer therapy. Mol Pharm 19(8): 2877–2887
14. Jerath G, Goyal R, Trivedi V, Santhoshkumar TR, Ramakrishnan V (2020) Conformationally constrained peptides for drug delivery. J Pept Sci 26(4–5):e3244
15. Goyal R, Jerath G, Chandrasekharan A, Kumar TRS, Ramakrishnan V (2020) Peptide-based delivery vectors with pre-defined geometrical locks. RSC Med Chem 11(11):1303–1313
16. Richard JP, Melikov K, Vives E, Ramos C, Verbeure B, Gait MJ, Chernomordik LV, Lebleu B (2003) Cell-penetrating peptides. A reevaluation of the mechanism of cellular uptake. J Biol Chem 278(1):585–590

Chapter 10

Cellular Uptake of Peptides by Confocal Microscopy

Laxmi Raman Adil and Yvonne Christian

Abstract

Confocal microscopy is a very sensitive and powerful optical imaging technique in vivo and in vitro imaging. Live cell imaging using fluorescence microscopy techniques has revolutionized biology. It has given access to the observer to study protein structures and important cellular process.

Key words Confocal laser scanning microscopy, Fluorescence, Fluorophore, Cell-penetrating peptide

1 Introduction

Confocal microscopy is an important tool for investigating biological structures and mechanism of various processes. Fluorescence-based microscopy has numerous advantages over other microscopic techniques due to its higher sensitivity and specificity [1]. The contribution of light-based technologies was highlighted by the United Nations General Assembly by declaring the year 2015 as the international year of light [2]. When a molecule is excited with UV-visible light, it will go to the excited state from S_0 to S_n (states). Furthermore, it will undergo radiative and non-radiative relaxation to the ground state from S_1 to S_0 [3]. Imaging is considered "the eyes of science" as it allows visualization of various physical processes in biological systems. As new methods and tools based on fluorescence techniques are introduced, we can get in-depth information on cellular structures and functions. As a recognition of its impact in science, the 2014 Nobel prize was jointly awarded to Eric Betzig, Stefan Hell, and William E. Moerner for the development of super-resolved fluorescence microscopy [4] (Fig. 1).

The following are the advantages of fluorescence confocal microscopy technique [5–7]:

1. Fluorescence-based techniques are non-destructive and non-invasive.

Vibin Ramakrishnan (ed.), *Biophysical Characterization of Functional Peptides*, Springer Protocols Handbooks,
https://doi.org/10.1007/978-1-0716-3405-9_10,

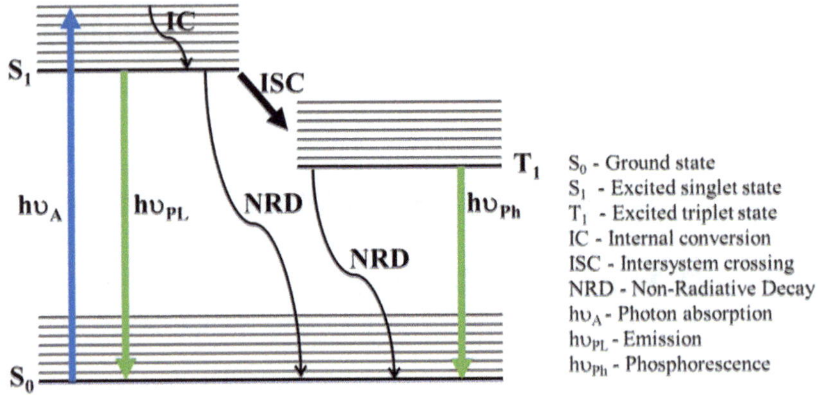

Fig. 1 Simplified Jablonski diagram [3]

2. Have better specificity and sensitivity.
3. Requires minimal pretreatment of sample and avoids physical sectioning.
4. Fluorescence of materials is highly dependent on their environment and therefore it is an excellent technique for monitoring cellular processes and biochemical reactions.

Confocal microscopy has advantage over light microscopy due to its ability to remove out-of-focus light owing to a pinhole aperture system. The pinhole allows passage of light only from a single focal plane [8, 9].

Cell-penetrating peptides (CPPs) have the ability to cross the cellular membrane. The cellular uptake and internalization of CPPs that are tagged with a fluorophore can be studied by various techniques such as fluorescence microscopy, flow cytometry, and fluorescence spectroscopy. Novel CPPs covalently tagged with the fluorophore 5(6)-carboxyfluorescein at the N-terminus were synthesized to study cellular uptake [10–12]. Tumor homing peptides having the ability to target tumor cells are utilized. Tumor homing peptide attached with a CPP and a fluorophore molecule has been shown to specifically internalize in cancerous cells to a much larger extent than non-cancerous cells [13, 14].

In this chapter we present a brief outline of using confocal laser scanning microscopy (CLSM) to visualize cellular uptake of a cell-penetrating peptide.

2 Materials

2.1 Equipment

1. 1300 Series Class II, Type A2 Biological Safety Cabinet (Thermo Fisher Scientific).
2. Galaxy 170 S CO_2 incubator (Eppendorf, New Brunswick).

3. 3K30 centrifuge (Sigma).
4. Laser scanning confocal microscope (Nikon A1R).

2.2 Reagents/Supplies

1. Cell lines: Triple negative breast cancer cell line MDA-MB-231 was purchased from ATCC.
2. Culture media: The cell line was cultured in DMEM media (Gibco) supplemented with 10% FBS and 1% penicillin (10,000 units/mL)–streptomycin (10,000 μg/mL).
3. 0.05% Trypsin-EDTA solution for cell dissociation (Gibco).
4. Phosphate-buffered saline (PBS), pH 7.4 (Gibco).
5. 0.4% Trypan Blue solution (Gibco).
6. Sterile 96-well black colored clear bottom tissue culture plate (Becton Dickinson).
7. 25 cm^2 rectangular canted neck cell culture flask with vent cap (Corning).
8. 5 mL polystyrene round-bottom tube (Falcon, Corning).
9. CPPs (de novo designed and TAT) were synthesized manually by solid-phase peptide synthesis and attached with the fluorophore 5(6)-carboxyfluorescein at the N-terminal.

3 Method

1. MDA-MB-231 cells are cultured in T-25 flasks in DMEM media containing 1 mg/mL penicillin-streptomycin antibiotic solution and 10% FBS. The flasks are maintained in an incubator at 37 °C with 5% CO_2 and 95% humidity.
2. Cells are detached using 0.05% Trypsin-EDTA solution. Cells were counted by staining with 0.4% Trypan Blue solution in a hemocytometer.
3. 10,000 cells per well are seeded in a 96-well plate and placed in an incubator at 37 °C for 24 h.
4. The next day, discard the media and wash the well with PBS once.
5. 5 μM of the novel CPP was prepared in 100 μL serum-free media. As a positive control 5 μM TAT-CF peptide is used.
6. The cells were treated with 5 μM peptide and the plate was incubated for 3 h at 37 °C.
7. After 3 h cells are washed thrice with PBS (*see* **Note 3**).
8. 5 μg/mL Hoechst 33342 dye is added for 5 min and the cells are incubated in a 37 °C incubator for 5 min.
9. Proceed for visualization using a confocal laser scanning microscope.

Table 1
Excitation and emission wavelengths of fluorescent dyes used for imaging

S. no	Fluorescent dye	Excitation (λ_{max})	Emission (λ_{max})
1	Peptide conjugated with 5(6)-carboxyfluorescein	489	517
2	Hoechst 33342 dye	346	460

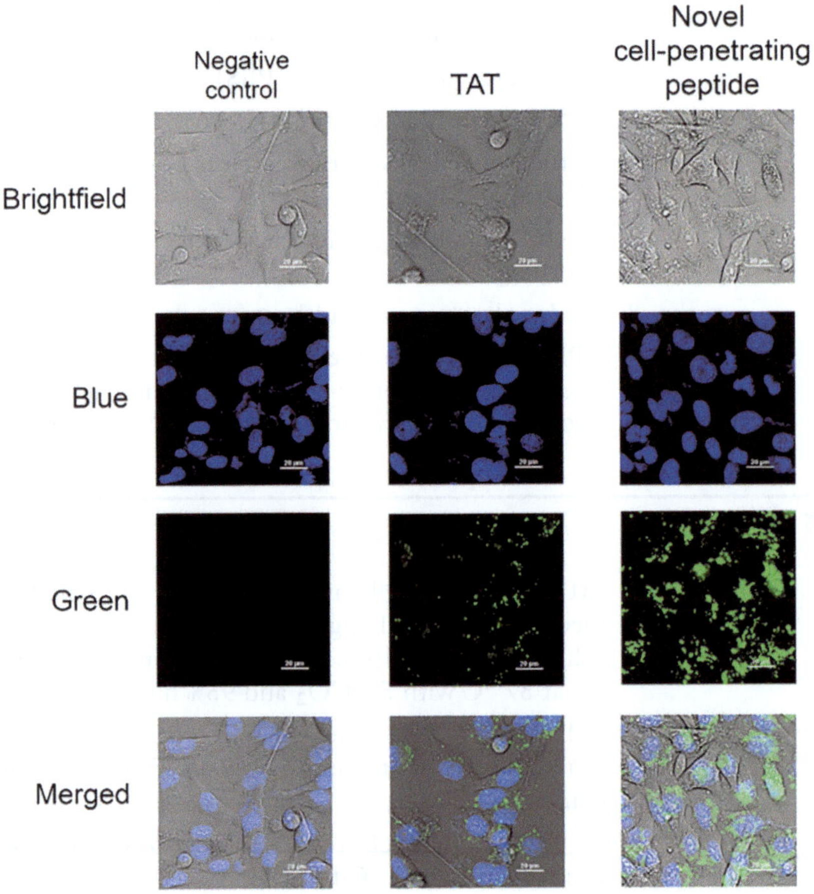

Fig. 2 Cellular uptake confocal laser scanning microscope. MDA-MB-231 cells were treated with a de novo designed peptide and TAT-CF peptide for 3 h and stained with Hoechst 33342 for nuclear staining. Blue represents nuclei staining by Hoechst 33342, green represents peptide-CF conjugate, and merged image represents the overlay. Scale bar corresponds to 20 μm

Table 1 represents the maximum excitation and emission wavelength for carboxyfluorescein-tagged peptides and Hoechst 33342 stain.

The panel (Fig. 2) describes the qualitative uptake of a de novo designed CPP and TAT in MDA-MB-231 cells. Three sets of images were taken—one with a de novo designed peptide, one with positive control (TAT-CF), and one without any peptide. The image shows that the novel designed CPP displays internalization to a greater extent than TAT peptide in cancerous MDA-MB-231 cells.

4 Notes

1. Always wear individual protection equipment such as lab coat, gloves, and safety goggles while handling chemicals and biological samples.
2. Operators need to be well-trained in order to utilize a confocal microscope.
3. Cells must be washed well in order to remove background fluorescence.
4. Source of confocal microscope emits UV light, therefore must be careful to not look directly at the lamp.
5. CLSM machine with its lasers must be switched ON 15 min prior to imaging.

References

1. Rigby PJ, Goldie RG (1999) Confocal microscopy in biomedical research. Croat Med J 40(3):346–352
2. Mullin AS, Fourkas JT (2015) The international year of light and the chemistry classroom. J Phys Chem Lett 6(19):3882–3883
3. Lakowicz JR (2006) Principles of fluorescence spectroscopy, 3rd edn. Springer, Berlin
4. Möckl L, Lamb DC, Bräuchle C (2014) Super-resolved fluorescence microscopy: Nobel prize in chemistry 2014 for Eric Betzig, Stefan hell, and William E. Moerner. Angew Chem Int Ed Engl 53(51):13972–13977
5. Swain BC, Das AK, Rout J, Biswas S, Tripathy U (2022) Fluorescence correlation spectroscopy: a highly sensitive tool for probing intracellular molecular dynamics and disease diagnosis. In: Sahoo H (ed) Optical spectroscopic and microscopic techniques. Springer, Singapore
6. Pulgarín JA, Molina AA, Boras N (2021) Use of total fluorescence spectroscopy for the highly sensitive simultaneous determination of fluoroquinolones in rabbit plasma. J Anal Chem 76:733–741
7. Reynaud K, Nogueira D, Cortvrindt R, Kurzawa R, Smitz J (2001) Confocal microscopy: principles and applications to the field of reproductive biology. Folia Histochem Cytobiol 39(2):75–85
8. Elliott AD (2020) Confocal microscopy: principles and modern practices. Curr Protoc Cytom 92(1):e68
9. Nwaneshiudu A, Kuschal C, Sakamoto FH, Anderson RR, Schwarzenberger K, Young RC (2012) Introduction to confocal microscopy. J Invest Dermatol 132(12):e3
10. Jerath G, Goyal R, Trivedi V, Santhoshkumar TR, Ramakrishnan V (2019) Syndiotactic peptides for targeted delivery. Acta Biomater 87: 130–139
11. Jerath G, Goyal R, Trivedi V, Santhoshkumar TR, Ramakrishnan V (2020) Conformationally constrained peptides for drug delivery. J Pept Sci 26(4–5):e3244
12. Jerath G, Darvin P, Christian Y, Trivedi V, Kumar TRS, Ramakrishnan V (2022) Delivery of small molecules by syndiotactic peptides for breast cancer therapy. Mol Pharm 19(8): 2877–2887

13. Goyal R, Jerath G, Chandrasekharan A, Christian Y, Kumar TRS, Ramakrishnan V (2022) Molecular hybridization combining tumor homing and penetrating peptide domains for cellular targeting. Drug Deliv Transl Res 12(5):1285–1292
14. Goyal R, Jerath G, Chandrasekharan A, Kumar TRS, Ramakrishnan V (2020) Peptide-based delivery vectors with pre-defined geometrical locks. RSC Med Chem 11(11):1303–1313

Chapter 11

Penetration in 3D Tumorospheres

Ruchika Goyal and Vibin Ramakrishnan

Abstract

Cellular 3D tumorospheres or spheroids are improved models for performing biological assays related to cell differentiation, migration, drug testing, targeting, and penetration. They are in great demand in the pharmaceutical industry as they provide insights and indications during the pre-clinical evaluation phases. They mimic tumor architecture in terms of cellular organization, metabolite gradient, and cellular interactions. In our study, we have developed MDA-MB-231 breast cancer spheroids to evaluate the penetration of our designed peptide molecules in tumor-like constructs using confocal imaging.

Key words Tumorospheres, Spheroids, Liquid overlay technique, Peptide penetration

1 Introduction

Tumorospheres (spheroids) are the three-dimensional (3D) growth of cells in in vitro conditions. In general, they are tightly bound cells grown in a spherical shape, representing the 3D cell culture models that closely mimic the complex microenvironment of different tissues. Spheroids are preferred to replicate the characteristics of solid tumors due to their ability to show metabolic gradients of nutrients, oxygen, carbon dioxide, etc., similar to tumor vasculatures [1–3]. These gradients led to the formation of outer, intermediate, and inner layers of cells (Fig. 1a). The outer layer is composed of proliferative cells, the intermediate has quiescence cells, and the inner layer is comprised of necrotic cells due to hypoxia [1]. Many drugs are screened based on their distribution patterns and cytotoxicity profiles in tumorospheres, before going to the pre-clinical studies in animal models. Their easy handling in in vitro conditions makes them a preferred choice to conduct drug-screening assays.

These 3D cellular spheroids are cultured using scaffold-based and scaffold-free methods. Artificial matrices made of non-adhesive polymers like hyaluronic acid, agarose, and other biomaterials are

Vibin Ramakrishnan (ed.), *Biophysical Characterization of Functional Peptides*, Springer Protocols Handbooks,
https://doi.org/10.1007/978-1-0716-3405-9_11,

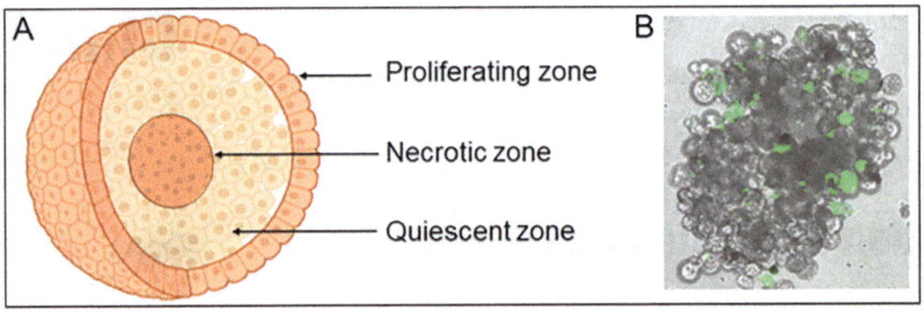

Fig. 1 Peptide penetration in spheroid. (**a**) Structure of a multicellular tumorosphere having three distinct zones of proliferating cells, quiescent cells, and necrotic cells. (**b**) A confocal plane of tumorosphere captured in both bright field and green fluorescence channels after peptide treatment, representing the penetration of carboxyfluorescein tagged peptide up to that plane [4]

used as a scaffold for spheroid formations. These can be hydrogels, beads, fibers, or sponges having porosity and mechanical traits similar to extracellular matrix of tumors [1, 5]. When cells are seeded on the matrix, they migrate in the matrix pores and form spheroids or micro-tissue like architecture [6]. In contrast, scaffold-free methods include liquid overlay technique, agitation-based methods, hanging drop technique, and micro-based techniques [2, 7, 8]. In our studies, we have employed the liquid overlay technique (LOT) to evaluate the penetration of our designed peptides in MDA-MB-231 breast cancer spheroids [4, 9].

2 Materials

2.1 Equipment

1. Biosafety cabinet (Micro-Filt India, MFI BIO 3*2) for cell culture work.
2. CO_2 incubator (Thermo Fisher Scientific, Forma Steri-Cycle) maintained at 37 °C with 5% CO_2 in a humidified atmosphere.
3. Refrigerated centrifuge (Eppendorf, Centrifuge 5810 R) for centrifugation purposes.
4. Confocal Laser Scanning Microscope (NikonA1R, Z-scan model) for imaging the penetration of tagged peptides.
5. Inverted Light microscope (Nikon eclipse Ti) to image cells and the growth of spheroids.
6. Single channel pipettes of different measurements to dispense solutions (Thermo Fisher Scientific).

2.2 Reagents/ Supplies

1. Cell line: MDA-MB-231 breast cancer cell line (purchased from NCI, USA).
2. Cell culture medium:

   (a) DMEM supplemented with 10% fetal bovine serum (FBS), 1% antibiotic cocktail (25 μg/mL amphotericin B, 10,000 units/mL Penicillin, and 10 mg/mL streptomycin), and sodium bicarbonate ($NaHCO_3$) salt.
   (b) Cancer stem cell (CSC) Media (Gibco, USA).
3. Phosphate buffered saline, pH 7.4 (PBS, Sigma).
4. Trypsin-EDTA (0.05%, Thermo Fisher Scientific) for detachment of cells.
5. Carboxyfluorescein tagged peptides to be tested for penetration in spheroids.
6. 1% Agarose solution for coating cell culture dishes.
7. 60-mm cell culture dishes and 96-well glass bottom low attachment tissue culture plates for high content imaging (Corning).
8. RIA vials.

3 Methods

3.1 Cell Culture

1. Thaw the freezing stocks (1 mL) of MDA-MB-231 cell line in 9 mL DMEM in 15 mL falcon tubes.
2. Centrifuge the cultures at 900 rpm for 5 min.
3. Decant the supernatant slowly and re-suspend the pellets in fresh media.
4. Seed cells in T-25 flasks having at least 5 mL of DMEM.
5. Incubate the flasks in CO_2 incubator at 37 °C overnight.

3.2 Tumorosphere Culture Using LOT

1. Trypsinize the overnight grown cells and seed in low density in a 60 mm agarose (1%) coated dish.
2. Grow the cells for 7–10 days at 37 °C under water-saturated conditions with 5% CO_2.
3. Keep checking for contamination, numbers, and morphology of tumorospheres regularly using a light microscope during the incubation period.
4. Perform washing with 1X PBS and media change gently without disturbing the tumorospheres, only if required.

3.3 Treatment with Test Molecules

1. Collect the well-formed tumorospheres in the separate wells of low attachment 96-well plates by cutting the tip wide.
2. Give peptide treatment for 4 h at the optimum concentration.
3. Wash the spheroids with 1× PBS gently thrice to remove excess peptide solution.
4. Perform the imaging of treated spheroids under a confocal microscope in Z-scan mode to capture each plane and construct a 3D image.

3.3.1 Expected Outcome

The images of treated tumorospheres were captured in both bright field mode and green fluorescence channel (Fig. 1b). This has been elaborately explained in our earlier published studies [4, 9]. In addition, a volume projection view with depth-coded color was also recorded to verify the peptide penetration [4, 9]. Peptide based delivery vehicles and their potential utility has regained focus in recent times, especially in nanomedicine [10–12].

4 Notes

1. Keep a regular check on contamination while growing spheroids with minimal disturbance.
2. Use freshly prepared peptide solution for staining.
3. Prefer to use a clear cell culture medium (without phenol red indicator) for imaging.
4. Use gentle pipetting while handling or transferring tumorospheres.
5. To avoid cell clumping, prefer to use a sterile cell strainer during seeding cells for the formation of tumorospheres.

References

1. Costa EC, Moreira AF, de Melo-Diogo D, Gaspar VM, Carvalho MP, Correia IJ (2016) 3D tumor spheroids: an overview on the tools and techniques used for their analysis. Biotechnol Adv 34(8):1427–1441
2. Mehta G, Hsiao AY, Ingram M, Luker GD, Takayama S (2012) Opportunities and challenges for use of tumor spheroids as models to test drug delivery and efficacy. J Control Release 164(2):192–204
3. Friedrich J, Seidel C, Ebner R, Kunz-Schughart LA (2009) Spheroid-based drug screen: considerations and practical approach. Nat Protoc 4(3):309–324
4. Goyal R, Jerath G, Chandrasekharan A, Christian Y, Kumar TRS, Ramakrishnan V (2022) Molecular hybridization combining tumor homing and penetrating peptide domains for cellular targeting. Drug Deliv Transl Res 12(5):1285–1292
5. Carvalho MP, Costa EC, Miguel SP, Correia IJ (2016) Tumor spheroid assembly on hyaluronic acid-based structures: a review. Carbohydr Polym 150:139–148
6. Rimann M, Graf-Hausner U (2012) Synthetic 3D multicellular systems for drug development. Curr Opin Biotechnol 23(5):803–809
7. Lin RZ, Chang HY (2008) Recent advances in three-dimensional multicellular spheroid culture for biomedical research. Biotechnol J 3(9–10):1172–1184
8. Breslin S, O'Driscoll L (2013) Three-dimensional cell culture: the missing link in drug discovery. Drug Discov Today 18(5–6): 240–249
9. Goyal R, Jerath G, Akhil R, Chandrasekharan A, Puppala ER, Ponneganti S, Sarma A, Naidu VGM, Santhoshkumar TR, Ramakrishnan V (2021) Geometry encoded functional programming of tumor homing peptides for targeted drug delivery. J Control Release 333:16–27
10. Jerath G, Darvin P, Christian Y, Trivedi V, Kumar TRS, Ramakrishnan V (2022) Delivery of small molecules by Syndiotactic peptides for breast cancer therapy. Mol Pharm 19(8): 2877–2887
11. Jerath G, Goyal R, Trivedi V, Santhoshkumar TR, Ramakrishnan V (2020) Conformationally constrained peptides for drug delivery. J Pept Sci 26(4–5):e3244
12. Jerath G, Goyal R, Trivedi V, Santhoshkumar TR, Ramakrishnan V (2019) Syndiotactic peptides for targeted delivery. Acta Biomater 87: 130–139

Chapter 12

TMRM Cytotoxicity Assay

Ruchika Goyal and Vibin Ramakrishnan

Abstract

Mitochondrial functioning is a vital parameter to assess cellular health. It produces energy as ATP molecules by oxidative phosphorylation through an electrochemical gradient between the mitochondrial membranes. The impairment of membrane potential during cell damage led to its depolarization and loss of cellular integrity. Potentiometric dyes like tetramethylrhodamine methyl ester (TMRM) detect cellular toxicity by utilizing this mitochondrial membrane potential as an indicator.

Key words TMRM, Mitochondria, Cytotoxicity, Mitochondrial membrane potential, Potentiometric dye assay

1 Introduction

Mitochondria, the powerhouse of cells, are the vital subcellular compartment for energy generation in the living cells. It produces energy as adenosine triphosphate (ATP) through the process of oxidative phosphorylation through the electron transport chain (ETC). In many cell-related studies, the soundness of cellular health is estimated by evaluating the functional activity of mitochondria [1]. The therapeutic safety or targeted cytotoxicity of any drug is assessed by estimating the cell death after the drug treatment. Cell death can happen by necrosis or apoptosis. Necrosis is the sudden death of cells leading to inflammation, while apoptosis is programmed cell death or cell suicide, where the cells die progressively. The apoptotic cell death can be assessed by TMRM assay. TMRM is a fluorescent potentiometric dye that localizes in the mitochondria of healthy cells in a voltage-dependent manner [2, 3]. TMRM has excitation and emission maxima at 548 nm and 574 nm, respectively. Mitochondrial impairment in cell death conditions causes depolarization of mitochondrial membrane potential (Ψ_m). Consequently, it opens the mitochondrial permeability transition pore (PTP), which promotes caspase-dependent

Vibin Ramakrishnan (ed.), *Biophysical Characterization of Functional Peptides*, Springer Protocols Handbooks, https://doi.org/10.1007/978-1-0716-3405-9_12,

cell death by the leakage of intermembrane proteins like cytochrome c [4]. Upon cellular damage, the collapse of mitochondrial membrane potential releases TMRM in the cytosol that drops the TMRM fluorescence signal, an indication of cellular death (Fig. 1a).

2 Materials

2.1 Equipment

1. Biosafety cabinet (Micro-Filt India, MFI BIO 3*2) for mammalian cell culture-related experiments.
2. CO_2 incubator (Thermo Fisher Scientific, Forma Steri-Cycle) maintained at 37 °C with 5% CO_2 and humidified atmosphere.
3. Refrigerated centrifuge (Eppendorf, Centrifuge 5810 R) for centrifugation.
4. High throughput Pathway Bioimager 435 (BD Biosciences, USA) for imaging of cells.
5. Multichannel and single channel pipettes of different measurements for dispensing solutions (Tarsons).

2.2 Reagents/Supplies

1. Cell lines: MDA-MB-231 breast cancer cell line (purchased from NCI, USA), MCF-10A non-cancerous mammary epithelial cell line (Sigma).
2. Cell culture medium: Dulbecco's Modified Eagle Media (DMEM)-Phenol red and serum-free but supplemented with 1% antibiotics (25 μg/mL amphotericin B, 10,000 units/mL penicillin, and 10 mg/mL streptomycin), and sodium bicarbonate ($NaHCO_3$) salt for TMRM assay. Supplemented with 10% fetal bovine serum, 1% antibiotic cocktail (as mentioned above), and $NaHCO_3$ salt.

   Mammary Epithelial Cell Growth Medium (MEGM, Lonza, USA) supplemented with insulin, BPE, hEGF, hydrocortisone, and GA-1000.
3. Phosphate buffered saline, pH 7.4 (PBS, Sigma).
4. Trypsin-EDTA (0.05%, Thermo Fisher Scientific) for cells detachment.
5. Tetramethylrhodamine methyl ester perchlorate (TMRM) dye (Sigma-Aldrich), a mitochondrial membrane potential indicator.
6. Hoechst 33342 (Sigma) for nucleus staining.
7. 96-well glass bottom tissue culture plates for high content imaging (Corning).

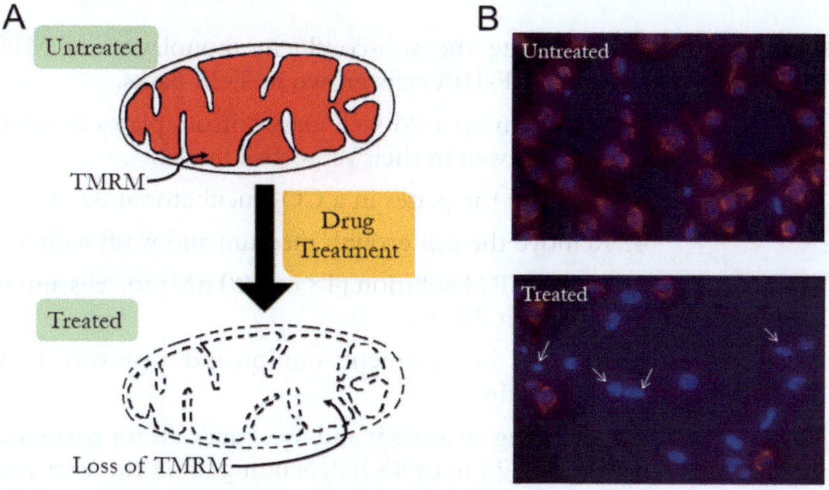

Fig. 1 TMRM assay principle. (**a**) Schematic diagram to show TMRM loss after drug treatment. (**b**) Representative results before and after drug treatment. The pointed arrows in the treated cells condition indicate TMRM loss. Nuclei are stained with Hoechst 33342 to indicate the presence of cells. Also, in addition to TMRM loss, the enhanced blue fluorescence after drug treatment shows chromatin condition, another hallmark of apoptotic cell death [5]

3 Methods

3.1 Cell Culture

1. Thaw the freezing stocks (1 mL) of MDA-MB-231 and MCF-10A cell lines in 9 mL DMEM and MEGM media, respectively, in 15 mL centrifuge tubes.
2. Centrifuge the cultures at 900 rpm for 5 min.
3. Decant the supernatant slowly and resuspend the pellets in fresh media.
4. Seed cells in T-25 flasks having at least 5 mL of respective media.
5. Incubate the flasks in a CO_2 incubator at 37 °C overnight.

3.2 Preparation of Stock Solutions

TMRM is available in both powder and liquid reagents by different companies. Prepare 100 μM TMRM solution in DMSO as stock solution. This can be stored for 6 months at −5 °C to −30 °C in amber microcentrifuge tubes.

3.3 Preparation of Working Solutions

100 μM TMRM stock solution is 1000× concentrated. The working concentration lies in the range of 20 nM to 250 nM, depending on the type of cell assay. For instance, we prepared a 100 nM working solution (1×) to perform this experiment [5]. We added 10 μL of 100 μM stock solution to 10 mL of PBS or cell growth medium to make it 1×. For best results, prepare a fresh staining solution every time before use.

3.4 TMRM Assay Protocol

1. Trypsinize the sub-confluent monolayers of MDA-MB-231 and MCF-10A cells grown in T-25 flasks.
2. Seed them in a 96-well glass bottom plates at a density of 10^4 cells per well in their respective media.
3. Incubate the plates in a CO_2 incubator at 37 °C overnight.
4. Remove the cell growth medium and wash with 1× PBS.
5. Add TMRM solution (1× or 100 nM) to cells and incubate for 30 min at 37 °C.
6. Remove the treatment solution and wash with 1× PBS for the best results.
7. Give drug treatment and incubate cells for particular durations like 6 h, 24 h, or 48 h depending upon the assay requirements.
8. After incubation, capture images of each well of the 96-well plate by high-throughput Pathway Bioimager or by any fluorescence microscope in red channel (TRITC/RFP filter) settings (Fig. 1b).
9. Analyze the data by checking TMRM distribution and signal quantification.
10. The cells undergoing apoptotic cell death after drug treatment show TMRM loss, which is detectable by direct imaging (Fig. 1b) and can be quantified (manually or by software like image J) for quantitative data representation [5–7].

4 Notes

1. Nucleus staining to visualize chromatin condensation (Fig. 1b) as an apoptotic marker is optional in this assay.
2. Use freshly prepared working solution.
3. Prefer to use a clear cell culture medium, i.e., without phenol red for imaging.
4. After treatment, cells can also be analyzed by flow cytometry. Laser with excitation at 488 nm and emission filters with 570 ± 10 nm can be used for detection. For such experiments, use 24-well plates to culture cells.

References

1. Rabelo ACS (2021) Chapter 17—Baccharis trimera phytochemicals and the mitochondria. In: de Oliveira MR (ed) Mitochondrial physiology and vegetal molecules. Academic Press, New York, pp 377–393
2. Scaduto RC Jr, Grotyohann LW (1999) Measurement of mitochondrial membrane potential using fluorescent rhodamine derivatives. Biophys J 76(1 Pt 1):469–477
3. Creed S, McKenzie M (2019) Measurement of mitochondrial membrane potential with the

fluorescent dye Tetramethylrhodamine methyl Ester (TMRM). Methods Mol Biol 1928:69–76

4. Budd SL, Tenneti L, Lishnak T, Lipton SA (2000) Mitochondrial and extramitochondrial apoptotic signaling pathways in cerebrocortical neurons. Proc Natl Acad Sci U S A 97(11): 6161–6166
5. Goyal R, Jerath G, Akhil R, Chandrasekharan A, Puppala ER, Ponneganti S, Sarma A, Naidu VGM, Santhoshkumar TR, Ramakrishnan V (2021) Geometry encoded functional programming of tumor homing peptides for targeted drug delivery. J Control Release 333:16–27
6. Goyal R, Jerath G, Chandrasekharan A, Kumar TRS, Ramakrishnan V (2020) Peptide-based delivery vectors with pre-defined geometrical locks. RSC Med Chem 11(11):1303–1313
7. Goyal R, Jerath G, Chandrasekharan A, Christian Y, Kumar TRS, Ramakrishnan V (2022) Molecular hybridization combining tumor homing and penetrating peptide domains for cellular targeting. Drug Deliv Transl Res 12(5):1285–1292

Chapter 13

Cell Death by Apoptosis

Ruchika Goyal, Khushboo Bhagat, and Vibin Ramakrishnan

Abstract

Apoptosis is a phenomenon involving multiple sequential molecular steps that lead to cell death. The intrinsic pathway and extrinsic pathway are two major pathways of apoptosis. They are executed with the help of different caspases. Caspase-3 is an executioner caspase that leads to irreversible cell death by apoptosis. This protocol uses a ratiometric fluorescence resonance energy transfer (FRET)-based assay to confirm cell apoptosis after drug treatment.

Key words Apoptosis, Programmed cell death, Caspase-3, FRET, Ratiometric imaging

1 Introduction

The term apoptosis (in Greek, the prefix "apo" refers to "separation" and the suffix, "ptosis," translated as "falling off") is considered similar to the falling off leaves from trees. Apoptosis or programmed cell death (PCD) is a complex biological process that systemically allows cellular death and helps an organism to kill and take out unwanted cells [1, 2]. In 1972, John Kerr first described the ultrastructural changes that happened during apoptosis [3]. There are distinct physical and biochemical changes in the cytoplasm, nucleus, and plasma membrane of cells undergoing apoptosis. This includes cell shrinkage, detachment of cytoskeleton from the cell membrane, membrane blebbing, chromatin condensation and fragmentation by endonucleases, etc. (Fig. 1a) [4, 5]. The interleukin-1 beta-converting enzymes (ICE)-like proteases and caspases are the crucial proteins involved in the activation and execution of apoptosis [6, 7]. Among different caspases, caspase-3 is a crucial mediator of programmed cell death.

In this protocol, we described a fluorescence resonance energy transfer (FRET) based probe to detect caspase-3 activation in the drug-treated cells (Fig. 1b). FRET probe is constituted by linking cyan fluorescent protein (CFP) to a yellow fluorescent protein

Vibin Ramakrishnan (ed.), *Biophysical Characterization of Functional Peptides*, Springer Protocols Handbooks,
https://doi.org/10.1007/978-1-0716-3405-9_13,

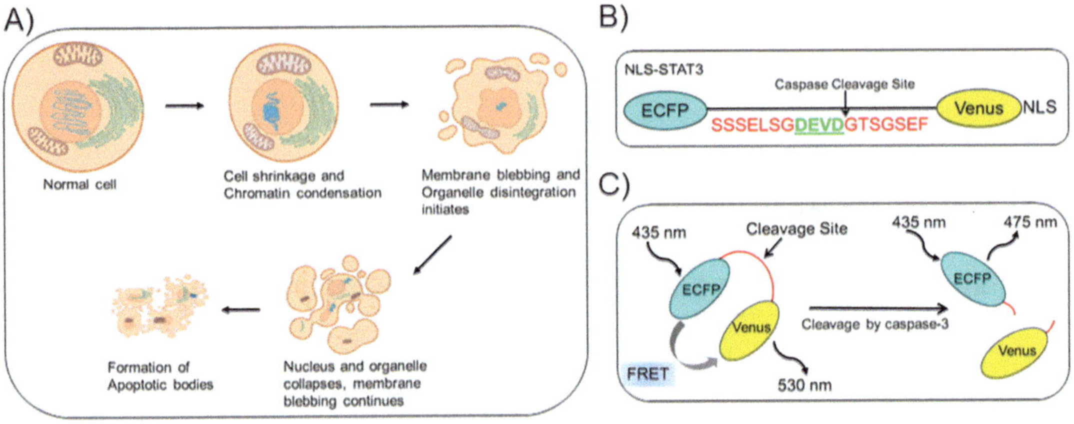

Fig. 1 Schematic illustration representing the phenomenon of apoptosis and its detection probe. (**a**) It shows the key features during apoptosis in a cell. (**b**) NLS-SCAT3 probe used in transformed MDA-MB-231 cell lines to detect caspase activation. (**c**) Caspase activation in cells led to FRET loss and hence, change in CFP/YFP ratio, which is captured using ratiometric imaging. (Created with BioRender.com)

(YFP) using a linker peptide sequence having a caspase-3 cleavage site (DEVD) [8, 9]. The peptide linker is short enough to bring CFP and YFP together, resulting in FRET.

After drug treatment, cells undergoing apoptosis show caspase-3 activation, which led to cleavage of the sensing probe at the DEVD site and, eventually, to FRET loss (Fig. 1c). The CFP/YFP ratio change is recorded through ratiometric imaging as a measure of caspase-3 activity or apoptotic cellular death [10–12]. This assay confirms apoptotic cell death in breast cancer MDA-MB-231 cells after drug treatment [10, 11].

2 Materials

2.1 Equipment

1. Biosafety cabinet (Micro-Filt India, MFI BIO 3*2) for performing mammalian cell culture experiments.
2. Inverted light microscope (Nikon eclipse Ti) to image cells.
3. CO_2 incubator (Thermo Fisher Scientific, Forma Steri-Cycle) maintained at 37 °C with 5% CO_2 and humidified atmosphere.
4. Refrigerated centrifuge (Eppendorf, Centrifuge 5810 R) for centrifugation.
5. High throughput Pathway Bioimager 435 (BD Biosciences, USA) for ratiometric imaging of cells.
6. Multichannel and single channel pipettes of different measurements for dispensing solutions (Tarsons).

2.2 Reagents/Supplies

1. Cell lines:
   (a) MDA-MB-231 breast cancer cell line (purchased from NCI, USA) has stable NLS SCAT-3 FRET probe expression.
2. Cell culture medium:
   (b) Dulbecco's Modified Eagle Media (DMEM)
   - Supplemented with 10% fetal bovine serum (FBS), 1% antibiotic cocktail (25 μg/mL amphotericin B, 10,000 units/mL penicillin, and 10 mg/mL streptomycin), and sodium bicarbonate ($NaHCO_3$) salt for cell culture
   - Phenol-red and serum-free but supplemented with a 1% antibiotic cocktail (as mentioned above) and sodium bicarbonate ($NaHCO_3$) salt for imaging after drug treatment
3. Phosphate buffered saline, pH 7.4 (PBS, Sigma)
4. Trypsin-EDTA (0.05%, Thermo Fisher Scientific) for detaching cells
5. Drug or molecules to be tested for apoptotic cytotoxicity in MDA-MB-231 cells
6. 96-well glass bottom tissue culture plates for high-content imaging (Corning)

3 Methods

3.1 Cell Culture

1. Thaw the freezing stocks (1 mL) of the MDA-MB-231 cell line having a stable expression of FRET-based caspase probe (DEVD) in 9 mL DMEM in 15 mL centrifuge tubes.
2. Centrifuge the cultures at 900 rpm for 5 min.
3. Decant the supernatant slowly and resuspend the pellets in fresh media.
4. Seed cells in T-25 flasks having at least 5 mL of DMEM supplemented with 10% FBS and 1% antibiotic cocktail.
5. Incubate the flasks in a CO_2 incubator at 37 °C overnight to grow cells as sub-confluent monolayers.

3.2 Assay Protocol

1. Trypsinize the sub-confluent monolayers of transformed MDA-MB-231 cells grown in T-25 flasks.
2. Seed them in 96-well glass bottom plates at a density of 10^4 cells per well in DMEM.
3. Incubate the plates in a CO_2 incubator at 37 °C for 24 h.
4. Remove the cell growth medium and wash with 1× PBS.

5. Give treatment of test molecules in phenol-red and serum-free DMEM at different concentrations and incubate for particular durations like 6 h, 24 h, or 48 h at 37 °C, depending upon the assay requirements.
6. After incubation, perform ratiometric imaging of CFP/YFP FRET by high-throughput pathway Bioimager 435 (BD Biosciences, USA) while exciting ECFP and collecting both enhanced cyan fluorescent protein (ECFP) and fluorescence resonance energy transfer (FRET) channels [10, 11].
7. Generate quantitative data by analyzing the captured images of treated cells [10, 11].

3.2.1 Expected Outcome

After treatment, the cells undergoing apoptotic cell death have activated caspase-3, which cleaves the CFP-YFP FRET probe at the DEVD site. Consequently, it increased cyanine fluorescence and, hence, CFP/YFP ratio. The ratio change is compared in treated cells with the cells without treatment by analyzing the captured images of cells in different channels [10, 11].

4 Notes

1. MDA-MB-231 cells having the stable expression of ECFP-VENUS SCAT 3 FRET probe were obtained from Dr. T. R. Santhoshkumar, Rajiv Gandhi Centre for Biotechnology, Thiruvananthapuram, Kerala, India.
2. Prefer to use a clear cell culture medium (without phenol red) for imaging.
3. After treatment, cells can also be analyzed manually or by software like Image J for quantitative data representation.
4. Always wear appropriate gloves and personal protective equipment while handling cancer cells.
5. Follow the recommended disposal procedure for disposing of biological waste.

References

1. Thompson CB (1995) Apoptosis in the pathogenesis and treatment of disease. Science 267: 1456–1462
2. Jacobson MD, Weil M, Raff MC (1997) Programmed cell death in animal development. Cell 88:347–354
3. Kerr JF, Wyllie AH, Currie AR (1972) Apoptosis: a basic biological phenomenon with wide-ranging implications in tissue kinetics. Br J Cancer 26:239–257
4. Saraste A, Pulkki K (2000) Morphologic and biochemical hallmarks of apoptosis. Cardiovasc Res 45:528–537
5. Johnson KL, Vaillant F, Alfons L (1996) Protein tyrosine kinase inhibitors prevent didemnin B-induced apoptosis in HL-60 cells. FEBS Lett 383:1–5
6. Nicholson DW, Thornberry NA (1997) Caspases: killer proteases. Trends Biochem Sci 22: 299–306

7. Caspases: Enemies Within | Science, https://www.science.org/doi/10.1126/science.281.5381.1312?url_ver=Z39.88-2003&rfr_id=ori:rid:crossref.org&rfr_dat=cr_pub%20%200pubmed
8. Xu X, Gerard AL, Huang BC et al (1998) Detection of programmed cell death using fluorescence energy transfer. Nucleic Acids Res 26:2034–2035
9. Mahajan NP, Harrison-Shostak DC, Michaux J et al (1999) Novel mutant green fluorescent protein protease substrates reveal the activation of specific caspases during apoptosis. Chem Biol 6:401–409
10. Goyal R, Jerath G, Akhil R et al (2021) Geometry encoded functional programming of tumor homing peptides for targeted drug delivery. J Control Release 333:16–27
11. Goyal R, Jerath G, Chandrasekharan A et al (2020) Peptide-based delivery vectors with pre-defined geometrical locks. RSC Med Chem 11:1303–1313
12. Joseph J, Seervi M, Sobhan PK et al (2011) High throughput ratio imaging to profile caspase activity: potential application in multiparameter high content apoptosis analysis and drug screening. PloS One 6:e20114

Chapter 14

Hemolytic Activity

Dikshita Hazarika, Yvonne Christian, and Vibin Ramakrishnan

Abstract

Hemolysis is the most commonly used method to evaluate toxicity for therapies administered intravenously. Red blood cells (RBCs) are separated using centrifugation and spectro-photometry is used to estimate the fraction of lysed RBCs in the prepared sample. A standard-curve is plotted with absorbance values, intended to estimate the amount of hemoglobin in the supernatant. The concentration of hemoglobin measured is then used to compare against the total hemoglobin concentration, so that level of peptide instigated hemolysis can be quantified. This protocol specifically outlines the procedure for quantifying the hemo-toxicity of cell-penetrating peptides.

Key words Hemolysis, Red blood cells (RBCs), Cell-penetrating peptides (CPPs), Toxicity, Hemoglobin

1 Introduction

Red blood cells include roughly 40–46% of the entire blood volume. They have a lifespan of 120 days in circulation. Hemoglobin is an important constituent protein in the RBCs and its function is to transfer oxygen from the lungs to different parts of the body. Hemolysis is the phenomenon of breakdown of red blood cells (RBCs) in circulation leading to the release of hemoglobin (Fig. 1). Such a condition in vivo causes physical conditions like paleness, jaundice, anemia, hypertension, and renal toxicity, which might turn into serious complications. The hemolytic activity is measured spectrophotometrically as the amount of hemoglobin released due to RBC lysis by incubation with various dilutions of peptides. Every clinical gadget or medicinal formulation that interacts with the bloodstream must be examined for hemolytic activity. This makes it one of the fundamental tests to evaluate toxicity. An extensively used indicator to estimate the toxicity levels of a peptide or protein is half maximum effective concentration (EC_{50} or HC_{50}). It may be defined as a measure of the concentration of a

Vibin Ramakrishnan (ed.), *Biophysical Characterization of Functional Peptides*, Springer Protocols Handbooks,
https://doi.org/10.1007/978-1-0716-3405-9_14,

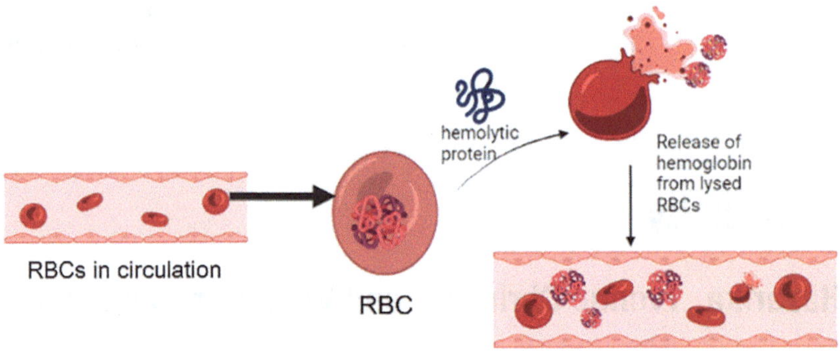

Fig. 1 Hemolysis of RBCs by hemolytic proteins. (Created with biorender.com)

drug (it may be an antibody or any toxicant) effective in generating 50% of the maximal response [1].

There are a few studies which compare hemolytic properties of various peptides on RBCs of various species. A study was conducted using synthetic anti-microbial peptide to compare the RBC lysis susceptibility of sheep, humans, and pigs [2]. Another example of comparison of hemolytic activity of cationic, amphipathic peptides on pigs, guinea pigs, and sheep RBCs showed variation due to a difference in phosphatidylcholine to sphingomyelin (PC/SM) ratio in the membrane [3]. A study by Helmerhorst et al. revealed that vulnerability of the RBCs to hemolysis increases along with the time of storage of the RBCs [4].

Apis mellifera (honey bee) produces a venom Melittin (26 residue amphipathic-peptide) which is considered as a "gold-standard positive control" as this peptide has a wide range of cell-lytic properties. Melittin is amphiphilic in nature with five basic side chain residues and has a bent α-helical conformation. It shows antibacterial and hemolytic properties [5]. It is also a cytotoxic agent toward many types of cancer cells but since it is non-specific toward tumors its use as an anti-cancer agent is limited. A particular study found that using a nanocarrier of RGD-modified PEG-lipids disk loaded with melittin significantly reduced the in vitro and in vivo hemolysis, thus increasing its efficacy as an anti-cancer agent [6]. For hemolytic assays, another reliable and comparatively cheaper substitute of melittin as a positive control is Triton-X, a non-ionic detergent.

Certain studies reveal that proteins with increasing hydrophobicity and proteins that are able to assume a stable amphipathic secondary structure usually have higher hemolytic activity [7]. For the antibacterial peptide indolicidin which is isolated from bovine neutrophils, the amino acid tryptophan was found to be an essential component causing hemolysis. In an experiment by Staubitz et al. when all five tryptophan residues in indolicidin (13 residue long peptide) were replaced with phenylalanine it had a positive influence on lysis of RBCs [8]. Furthermore, when all the tryptophan residues barring one were mutated to leucine in indolicidin, potent antibacterial activity and negligible hemolytic activity were reported [9]. Assessment of hemolytic activity to estimate the toxicity of cell penetrating, tumor homing and anti-microbial peptides have been discussed below. Cationic, helical amphipathic peptides (CHAPs) showing cell-penetrative properties and utilized to deliver methotrexate to breast and cervical cancer cells were tested for their hemolytic property toward mammalian RBCs. It was observed that CHAPs peptides showed negligible toxicity toward RBCs marking them non-toxic in the blood [10]. In another study, tumor homing peptides designed for targeted drug delivery to cancer cells were tested for their hemotoxicity to RBCs. For 100 μM of peptide the RBC lysis was seen to be less than 3% which indicated non-toxicity of the peptides [11]. Other examples of treatment of a peptide delivery vector consisting of tumor homing domain attached to a cell-penetrating domain showed insignificant RBC lysis after treatment for 2 h [12, 13]. Novel syndiotactic anti-microbial peptides against the organisms gentamicin resistant methicillin-resistant Staphylococcus aureus (MRSA), *E. coli,* and mycobacterium showed lower hemolytic activity against RBCs [14].

In this protocol, we outline a procedure for quantifying the hemolysis of a series of tumor homing peptides (RG101-RG108) against mammalian RBCs using Triton-X as a positive control [11].

2 Materials

- 1 X Phosphate buffered saline (PBS) buffer: 137 mM sodium chloride (NaCl), 2.7 mM potassium chloride, 10 mM disodium hydrogen phosphate (Na_2HPO_4), 1.8 mM potassium dihydrogen phosphate (KH_2PO_4) (*see* **Note 6**).
- 2 mL of blood from a healthy volunteer (*see* **Note 3).**
- Sterile falcon tubes.
- 0.5% Triton X-100 solution.
- Ethylenediaminetetraacetic acid (EDTA).
- Micro centrifuge tubes.
- Micro titer-plates consisting of 96-wells.
- Biorad-Microseal film.
- Autoclave.
- Incubator.
- Plate-centrifuge.
- Spectro-photometer.

3 Methods

3.1 Preparation of 10% Hematocrit

1. Collect 2 mL of whole blood from a healthy volunteer into a sterile falcon tube containing 3 mg of EDTA (*see* **Note 1**).
2. Mix the tube thoroughly for 5–10 min (*see* **Note 4**).
3. Centrifuge this tube at 1500 rpm (200 *g*) for 5 min. The supernatant containing the serum is then discarded gently.
4. Add 3 mL PBS buffer to the pellet remaining in the tube (*see* **Note 2 and 5**).
5. Centrifuge the mixture again at 1500 rpm (200 *g*) for 5 min and discard the supernatant.
6. Repeat **steps 4** and **5** multiple times until a clear supernatant is obtained.
7. After obtaining a clear supernatant, discard the last clear supernatant to get a pellet of 100% RBC.
8. 10% hematocrit of erythrocytes is then prepared by mixing 500 μL of 100% RBC in 4.5 mL of PBS (*see* **Note** 7).

3.2 Peptide Stock Preparation

Peptides were dissolved in sterile milli-Q water at a concentration of 200 μM.

3.3 Procedure

1. Dispense 50 μL of the prepared 10% hematocrit in sterile micro centrifuge tubes.
2. Add 50 μL of the prepared 200 μM of peptide stock solution into the micro centrifuge tubes and mix it well. This will make a total concentration of 100 μM peptide in 5% hematocrit.
3. As a positive control 0.5% of Triton X-100 is incubated in 5% hematocrit in a separate micro centrifuge tube (*see* **Note 8**).
4. Incubate the tubes for 2 h in an incubator at 37 °C under shaking conditions (*see* **Note 9**).
5. Centrifuge at 800 *g* for 5 min.
6. Collect the supernatant and dispense it into a 96-well plate. Each sample must be dispensed in triplicate wells.
7. Add 5% hematocrit separately in triplicate wells as a negative control (*see* **Note 10**).
8. Record the absorbance of the plate at 540 nm.

3.4 Result Interpretation

1. Individual mean of the negative and positive controls was calculated.
2. Normalization of the results was done in regard of the mean of the positive (100% lysis) and negative (0% lysis) controls.
3. Formula used to calculate % hemolysis is

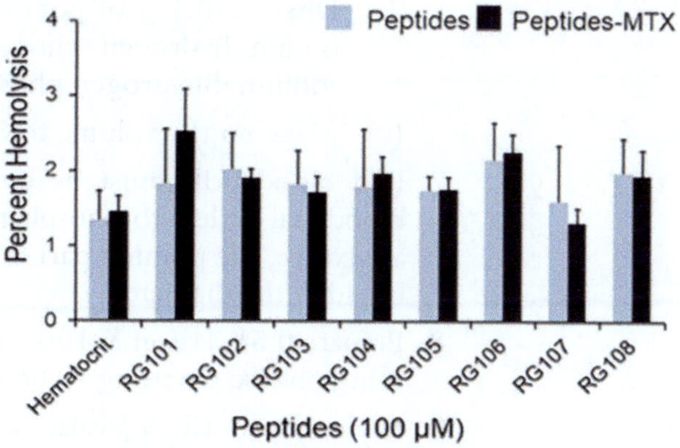

Fig. 2 Hemolytic assay of the peptides (RG101-RG108) and their methotrexate conjugates [11]. (Reprinted from 'Geometry encoded functional programming of tumor homing peptides for targeted drug delivery (2021), 333, Goyal R, Jerath G, Akhil R, Chandrasekharan A, Puppala ER, Ponneganti S, Sarma A, Naidu VGM, Santhoshkumar TR and Ramakrishnan V, 16-27, 2021, with permission from Elsevier)

$$\frac{\text{absobance of sample} - \text{mean absorbance of negative control}}{\text{mean absorbance of positive control} - \text{mean absorbance of negative control}} \times 100$$

The values obtained from the equation mentioned above in regard to % hemolysis were used to plot a graph (Fig. 2).

4. As observed from the graph, the percent of heme release from the designed peptide conjugates is <3% which is an indication of non-hemolytic nature of the peptides and peptide-MTX conjugates [11].

4 Notes

1. Always wear appropriate gloves and personal protective equipment while handling blood, peptide samples, and other reagents. Follow the recommended disposal procedure for disposing off the biological waste.
2. Vortexing leads to the damaging of erythrocytes and thus should be avoided.
3. Whole-blood must be kept at 4°C and should not be stored for more than 14 days.
4. Immediately after adding the blood, the centrifuge tube must be shaken vigorously or it might lead to clotting of the blood.
5. The PBS must be poured into the tube slowly through the walls of the tubes by slanting the tubes.
6. Preparation of phosphate buffer saline (PBS) (500 mL).
   (a) Add 4 g sodium chloride (NaCl) to 400 mL of ultrapure water.

(b) Dissolve 0.1 g of potassium chloride (KCl), 0.73ig disodium hydrogen phosphate (Na_2HPO_4), and 0.12 g sodium dihydrogen phosphate (NaH_2PO_4) in it.

(c) Make up the volume to 500 mL using ultrapure water.

7. Red blood cells must be dealt carefully. While shifting red blood cell-pellets to phosphate buffer saline in order to make a mixture, the pointed part of a 200 μL tip can be chopped to broaden the diameter.
8. Prepare 0.5% Triton X-100 in PBS by aliquoting the detergent using reverse pipetting technique.
9. Each tube must be sealed separately; else the sample might evaporate inside the incubator.
10. Each well in the 96-well plate must be labeled properly.

References

1. Munk JK, Ritz C, Fliedner FP, Frimodt-Moller N, Hansen PR (2014) Novel method to identify the optimal antimicrobial peptide in a combination matrix using anoplin as an example. Antimicrob Agents Chemother 58: 1063–1070
2. Dennison SR, Phoenix DA (2014) Susceptibility of sheep, human, and pig erythrocytes to hemolysis by the antimicrobial peptide Modelin 5. Eur Biophys J 43:423–423
3. Belokoneva OS, Villegas E, Corzo G, Dai L, Nakajima T (2003) The hemolytic activity of six arachnid cationic peptides is affected by the phosphatidylcholine-to-sphingomyelin ratio in lipid bilayers. Biochim Biophys Acta 1617:22–30
4. Helmerhorst EJ, Reijnders IM, Van 't Hof W, Veerman EC, Nieuw Amerongen AV (1999) A critical comparison of the hemolytic and fungicidal activities of cationic antimicrobial peptides. FEBS Lett 449(2-3):105–110
5. Yang L, Harroun TA, Weiss TM, Ding L, Huang HW (2001) Barrel-stave model or toroidal model? A case study on melittin pores. Biophys J 81:1475–1485
6. Neun BW, Dobrovolskaia MA (2011) Method for analysis of nanoparticle hemolytic properties in vitro. Methods Mol Biol 697:215–224
7. Tossi A, Sandri L, Giangaspero A (2000) Amphipathic, α-helical antimicrobial peptides. Biopolymers 55:4–30
8. Staubitz P, Peschel A, Nieuwenhuizen WF, Otto M, Götz F, Jung G, Jack RW (2001) Structure-function relationships in the tryptophan-rich, antimicrobial peptide indolicidin. J Pept Sci 7:552–564
9. Subbalakshmi C, Bikshapathy E, Sitaram N, Nagaraj R (2000) Antibacterial and hemolytic activities of single tryptophan analogs of indolicidin. Biochem Biophys Res Commun 274(3):714–716
10. Jerath G, Goyal R, Trivedi V, Santhoshkumar TR, Ramakrishnan V (2020) Conformationally constrained peptides for drug delivery. J Pept Sci 26(4-5):e3244
11. Goyal R, Jerath G, Akhil R, Chandrasekharan A, Puppala ER, Ponneganti S, Sarma A, Naidu VGM, Santhoshkumar TR, Ramakrishnan V (2021) Geometry encoded functional programming of tumor homing peptides for targeted drug delivery. J Control Release 333:16–27
12. Goyal R, Jerath G, Chandrasekharan A, Christian Y, Kumar TRS, Ramakrishnan V (2022) Molecular hybridization combining tumor homing and penetrating peptide domains for cellular targeting. Drug Deliv Transl Res 12(5):1285–1292
13. Goyal R, Jerath G, Chandrasekharan A, Kumar TRS, Ramakrishnan V (2020) Peptide-based delivery vectors with pre-defined geometrical locks. RSC Med Chem 11(11):1303–1313
14. Hazam PK, Phukan C, Akhil R, Singh A, Ramakrishnan V (2021) Antimicrobial effects of syndiotactic polypeptides. Sci Rep 11(1): 1823

Chapter 15

Antimicrobial Assay

Amay Sanjay Redkar and Vibin Ramakrishnan

Abstract

The rise of antimicrobial resistance has surged the investigation of alternative therapeutic solutions. One of the most appealing alternatives is antimicrobial peptides (AMPs). Microdilution is the most popular among many methods to estimate the efficacy of peptides as antimicrobial agents. It, alongside the microbial bactericidal concentration assay and field emission scanning electron microscopy (FE-SEM), can be used to verify whether a peptide has antimicrobial potential. This protocol outlines a step-by-step procedure to study the antimicrobial property of peptides using the microdilution method.

Key words Antimicrobial peptides, Microdilution, *Staphylococcus aureus*, *Escherichia coli*, Minimum inhibitory concentration, Minimum bactericidal concentration

1 Introduction

The rapid increase of antibiotic resistance has compelled the development of antimicrobials with improved new and non-conventional anti-infective therapies, including adjunctive or preventive tactics such as antibodies, probiotics, and vaccines [1, 2]. Since most of the drawbacks can be addressed, antimicrobial peptides (AMPs), are considered one of the most attractive alternatives to current antibiotics. Antimicrobial peptides are oligopeptides with a diverse range of amino acids, from two to more than a hundred [3]. While they have a broad spectrum of organisms that they can act upon, ranging from viruses to parasites, they have been extensively studied for their activity against bacteria. Incorporating D amino acids to diversify chain stereochemistry [4–7], fusion partners [8–10], multimeric design [11], and hybridization [12] are some of the different engineering strategies for improved antimicrobial peptide activity with significantly reduced hemolytic activity and greater bioavailability.

The primary assessment of any peptide, natural or designed, as an in vivo effective AMP, should always start with an in vitro antimicrobial assay. The outcome of antimicrobial assay can be

Vibin Ramakrishnan (ed.), *Biophysical Characterization of Functional Peptides*, Springer Protocols Handbooks,
https://doi.org/10.1007/978-1-0716-3405-9_15,

confounded due to various experimental conditions. Therefore, the procedures must be consistent for reproducibility between experiments and laboratories. While there are many suitable protocols to verify whether a particular molecule can inhibit the growth of bacteria, such as agar diffusion, turbidimetric and potentiometric assays [13], the broth microdilution growth inhibition assay is the most widely used method for peptides.

Dilution methods are often regarded as reference methods for in vitro antimicrobial assays and are also used to assess other susceptibility testing methods. In the broth dilution method, a defined number of colony-forming units (CFUs) are seeded within the liquid media with serially diluted concentrations (typically a two-fold dilution series, e.g., 128, 64, 32, 16 μg/mL) of the antimicrobial agent. Depending on the final volume of the testing media in each concentration reading point, the test is called macrodilution, final volume 2 mL, or microdilution, typically performed in a microtiter plate, with volumes less than 500 μL per well. In agar dilution, mid-log phase CFUs are seeded on thin gels with various concentrations of antimicrobial agents.

In the agar and broth dilution approaches, the minimum inhibitory concentration (MIC) is the antimicrobial agent's lowest concentration (in μg/mL or μM) that inhibits detectable growth under defined conditions. The typical range of MIC for AMPs is between 1 and 32 μg/mL, compared to conventional antibiotics, whose MIC is less than 1 μg/mL [14]. MIC can also classify bacteria based on their response as susceptible or resistant. Although MIC is determined directly in the case of agar dilution, it is interpreted from turbidity in broth dilution assay. Usually determined spectrophotometrically, the turbidity is read at 590 nm wavelength. The broth microdilution assay spans over 3 days. It involves primary culturing bacteria on the first day, AMP inculcated culture on the second day and reading the plates on the third day.

Much like MIC, the minimum bactericidal concentration (MBC) is the lowest effective concentration of the antimicrobial agent at which it is lethal rather than halting the growth. It can be performed in broth and agar, with a preference for the latter. Agar-based MBC is preferred because the number of CFUs can be counted directly on a plate rather than through indirect estimation. Field emission scanning electron microscopy (FE-SEM) can be used to image the perturbation of bacteria by the antimicrobial agent, especially AMPs, which have the propensity to target the bacterial cell membrane.

This chapter aims to outline the peptide antimicrobial assays. The step-by-step procedure, troubleshooting, and precautions are also discussed. This protocol is presented in a step-by-step layout and follows the guidelines of the Clinical and Laboratory Standards Institute (CLSI) and the European Committee on Antimicrobial Susceptibility Testing (EUCAST) with some modifications. The practices outlined in this protocol are reserved for only aerobic bacteria propagating effectively within 24 h in the nutrient broth per the CLSI and EUCAST guidelines [15].

2 Materials

2.1 Microorganisms

Escherichia coli MG1655 and *Staphylococcus aureus* NCTC 8530.

2.2 Media and Reagents

1. Nutrient Agar (NA): Dissolve 28 g in 1 L deionized water and autoclave at 121 ° C for 20 min.
2. Nutrient Broth (NB): Dissolve 13 g in 1 L deionized water and autoclave at 121 ° C for 20 min and store at room temperature.
3. Glycerol (50% v/v): Mix 10 mL glycerol with 10 mL deionized water and autoclave at 121 ° C for 20 min and store at 4 ° C.
4. Phosphate buffer saline (PBS): 10 mM with pH 7.4
5. Peptide solution(s), purified.

2.3 Instruments and Hardware

1. Inoculation loops.
2. Flasks.
3. Spectrophotometer.
4. Cuvettes.
5. Sterile 96-well microtiter plates.
6. Tips (1 mL, 200 μL, 20 μL).
7. Micropipette (20 μL, 200 μL, and 1000 μL).
8. Spreader.
9. Vortex mixer.
10. Incubator.
11. Microplate reader.

3 Methods

3.1 Glycerol Stock Preparation

1. Culture the bacteria on NA plates.
2. Transfer isolated colonies into NB using a sterile inoculation loop and incubate them at recommended growth temperature, typically at 37 ° C, for 16–18 h.
3. Dilute the broth culture 1:1 with 50% sterile glycerol and store as 1 mL aliquots at −80 ° C. Streak 10 μL inoculum on NA plates using a sterile inoculation loop for regular use, incubated overnight and then store at 4 ° C.

3.2 Primary Culture and Inoculum Preparation

1. Streak the bacterial isolates to be tested onto NA plates without any antibiotic to obtain single colonies (*see* **Note 1**).
2. Incubate the plates overnight for 18–24 h at recommended growth temperature, typically at 37 ° C (*see* **Note 2**).

3.3 Peptide Stock Solution Preparation

1. Dissolve 95% or more purity peptide (determined using HPLC) in 1 mL of an appropriate solvent, typically ultrapure water, to a final concentration of 512 μM and store it in ice (*see* **Note 3** and **Note 4**).

3.4 Minimum Inhibitory Concentration (MIC) Determination

1. Dispense 100 μL of NB into every well with each peptide row of the microtiter plate. Label the plate and lid.
2. Dispense 100 μL of peptide stock solutions (as prepared above) into the wells in the first column.
3. Mix the peptide and media into the wells in column one using a pipette set at 100 μL by pipetting up and down five times within the same well. Do not splash.
4. Draw 100 μL from the first column and add this to the second, diluting the second to a twofold dilution of the first column. Mix carefully using a pipette five times by pipetting up and down within the same well. Transfer 100 μL to the third column. Repeat this step up to the tenth column only. The same tips can be used for dilutions made across a single row. Figure 1 shows a typical plate configuration.
5. Remove 100 μL from the tenth column and discard it rather than putting it in the eleventh column.
6. Mix the primary culture prepared earlier by vortexing and diluting it to an appropriate inoculum size. 10^4–10^5 CFU/mL is an apt inoculum size for peptide-based antimicrobial microdilution assay (*see* **Note 5**).
7. Seed the bacterial culture into wells from the eleventh to the first column, in that order, using a pipette set to 5 μL. Do not seed CFUs in the twelfth column (media blank and control for the microtiter plate reader) (*see* **Note 6**).
8. Incubate the plates at 37 ° C or other desired temperatures.
9. Plate and spread the bacterial cultures on NA plates to ensure their purity. The spreader should be sterile before each use. Refer to Fig. 1 for troubleshooting based on growth.
10. Scan the plates with a microtiter plate reader when satisfactory growth is obtained (typically 18–36 h), measuring the absorbance at 590 nm. Use the twelfth column as the blank.
11. Determine the MIC by identifying the wells with over 90% inhibitory activity inferred from the absorbance at 590 nm wavelength relative to the eleventh well, i.e., the growth

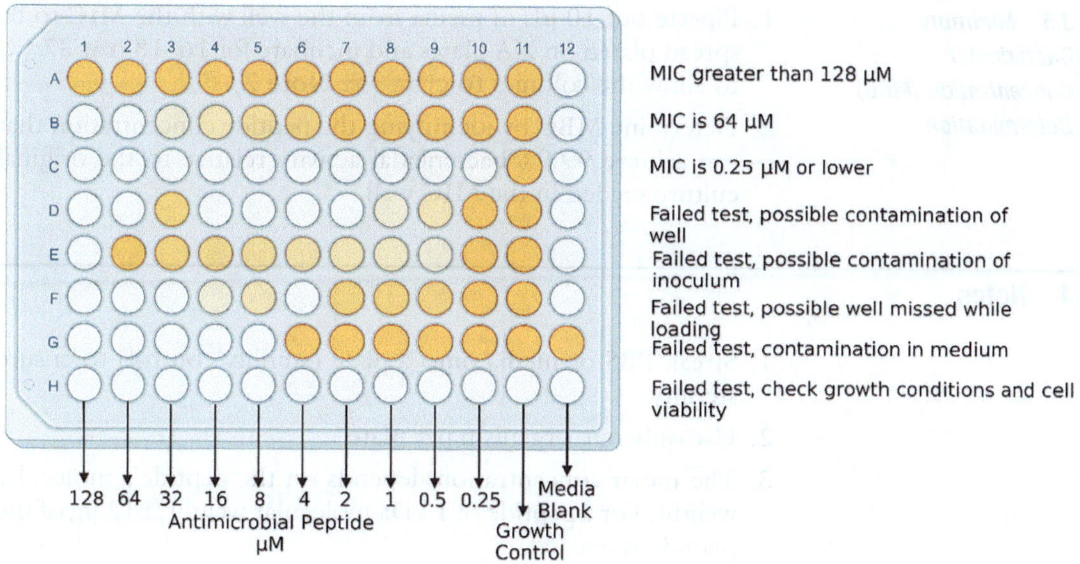

Fig. 1 Plate configuration and common inference of the readings based on the growth for troubleshooting (created with biorender.com)

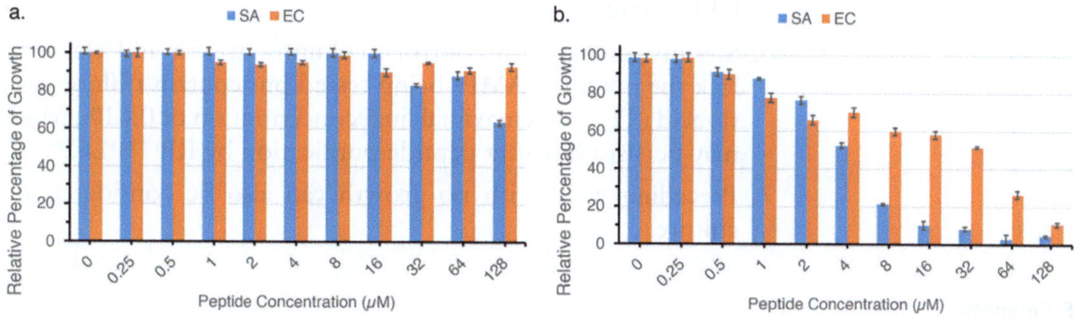

Fig. 2 Antimicrobial activity of a peptide using broth microdilution assay: (**a**) a peptide (AsSsKkKfF) with no significant activity from both *S. aureus* (SA) and *E. coli* (EC). (**b**) a peptide (kRklaArArLvA) with activity to *S. aureus* (SA) and *E. coli* (EC). Small letters indicate D amino acids

control. Figure 2 shows the result of a typical antimicrobial peptide microdilution assay. The relative percentage of growth in well can be calculated by using Eq. 1.

$$\%\mathrm{RG} = \frac{X - B}{G - B} \quad (1)$$

where %RG stands for relative percentage of growth

X stands for well in examination for growth.

B stands for media blank.

G stands for growth control.

3.5 Minimum Bactericidal Concentration (MBC) Determination

1. Pipette out 10 μL of media from the well with the MIC to be spread plated on NA plates and incubate for 16–18 h at 37 ° C to allow the colonies to grow (*see* **Note 7**).
2. Determine MBC by identifying the peptide concentration that can achieve 99.9% bactericidal activity relative to the original culture seeded in the MIC well.

4 Notes

1. Streak PBS or media onto a plate (sterility control) to ensure sterility.
2. Use only one organism per plate.
3. The molar concentration depends on the peptide's molecular weight. For a peptide of 1 kDa molecular weight, 512 μg of the peptide is needed.
4. Use only one peptide per three rows of a microtiter plate, executing the experiment in triplicates.
5. Once prepared, use inoculum within 30 min to preserve the CFU count.
6. Sterility and growth controls, columns twelfth and eleventh, lack the potential AMPs being tested but contain trifluoroacetic acid (TFA) at a maximal final concentration of 0.01%, which may persist even after peptide purification by RP-HPLC.
7. Preceding wells with no growth can also be considered for culture.

References

1. Mahlapuu M, Håkansson J, Ringstad L et al (2016) Antimicrobial peptides: an emerging category of therapeutic agents. Front Cell Infect Microbiol 6:194
2. Czaplewski L, Bax R, Clokie M et al (2016) Alternatives to antibiotics—a pipeline portfolio review. Lancet Infect Dis 16:239–251
3. Hazam PK, Akhil R, Jerath G et al (2019) Topological effects on the designability and bactericidal potency of antimicrobial peptides. Biophys Chem 248:1–8
4. Hazam PK, Goyal R, Ramakrishnan V (2019) Peptide based antimicrobials: design strategies and therapeutic potential. Prog Biophys Mol Biol 142:10–22
5. Hazam PK, Phukan C, Akhil R et al (2021) Antimicrobial effects of syndiotactic polypeptides. Sci Rep 11:1823
6. Hazam PK, Singh A, Chaudhary N et al (2019) Bactericidal potency and extended serum life of stereo-chemically engineered peptides against mycobacterium. Int J Pept Res Ther 25:465–472
7. Hazam PK, Jerath G, Kumar A et al (2017) Effect of tacticity-derived topological constraints in bactericidal peptides. Biochim Biophys Acta Biomembr 1859:1388–1395
8. Forde E, Devocelle M (2015) Pro-moieties of antimicrobial peptide prodrugs. Molecules 20: 1210–1227
9. Pane K, Durante L, Pizzo E et al (2016) Rational design of a carrier protein for the production of recombinant toxic peptides in Escherichia coli. PLoS One 11:e0146552
10. Yu H, Li H, Gao D et al (2015) Secretory production of antimicrobial peptides in Escherichia coli using the catalytic domain of a cellulase as fusion partner. J Biotechnol 214:77–82
11. Matthyssen T, Li W, Holden JA et al (2022) The potential of modified and multimeric

antimicrobial peptide materials as superbug killers. Front Chem 9:795433
12. Almaaytah A, Qaoud MT, Abualhaijaa A et al (2018) Hybridization and antibiotic synergism as a tool for reducing the cytotoxicity of antimicrobial peptides. Infect Drug Resist 11:835–847
13. Sabath LD (1976) The assay of antimicrobial compounds. Hum Pathol 7:287–295
14. Raheem N, Straus SK (2019) Mechanisms of action for antimicrobial peptides with antibacterial and antibiofilm functions. Front Microbiol 10:2866
15. Weinstein MP, Patel JB (2018) Methods for dilution antimicrobial susceptibility tests for bacteria that grow aerobically: M07-A11. Committee for Clinical Laboratory Standards, Wayne, PA

Chapter 16

Antimicrobial Assay with Peptide Based Hydrogels

Vivek Prakash and Vibin Ramakrishnan

Abstract

Recent developments in the peptide design strategies have contributed to the functional evolution of antimicrobial peptides as possible alternatives for conventional antibiotics. Among antimicrobial peptides, peptide based hydrogels have inherent antimicrobial potency. This is especially useful for topical applications to support wound healing. This chapter describes the step-by-step procedure to validate the antimicrobial property of peptide hydrogels.

Key words Antibacterial activity, Peptides, Hydrogel, FESEM imaging, ROS activity, Outer membrane permeabilization assay

1 Introduction

Bacterial infection remains a major concern in healthcare sector. Although conventional antibiotics play a big role to prevent bacterial infections, advancement in the development of new natural and synthetic solutions is needed from time to time. Bioinspired self-assembled peptide molecules have garnered significant attention in recent years for the development of drug delivery vehicles [1], quantum dots [2], tissue engineering [3], and tumor homing [4]. In last few decades peptide based antimicrobial molecules (AMP) have emerged as a promising option to counter antibiotic resistance [5]. Conventionally, antimicrobial agents are either encapsulated in gels for release or attached with template molecules [6]. In recent years, peptides having inherent antimicrobial properties have been developed and their efficacy as inherent antimicrobial hydrogels has been validated [7, 8]. Determination of antimicrobial potency of peptide hydrogels has different approaches compared to conventional antimicrobial assay, because defined peptide concentration is required to prepare peptide hydrogels. The main factors that play a crucial role are broth media, bacterial colonies

Vibin Ramakrishnan (ed.), *Biophysical Characterization of Functional Peptides*, Springer Protocols Handbooks,
https://doi.org/10.1007/978-1-0716-3405-9_16,

(CFU/mL), incubation temperature, contamination free workspace, and properly autoclaved accessories.

This chapter focuses on the peptide hydrogel based antimicrobial assay, common problems during assays, and few suggestions for troubleshooting.

2 Materials

1. For culture of bacterial colonies: Nutrient broth (N.B.) media, agar, 15 and 50 mL centrifuge tubes, micro-centrifuge tubes, spreader, tissue culture plates, 96 wells plates, conical flask, cotton plugs.
2. For antimicrobial assays: 1-N-phenylnaphthylamine (NPN), polymyxin B, acetone, 2,7-dichlorofluorescein diacetate (DCFH2-DA), phosphate buffer (pH 7.4).
3. For field emission scanning electron microscopic (FESEM) analysis: Phosphate buffer (pH 7.4), glutaraldehyde (25% in water), ethanol/water gradient (v/v): 20, 30, 50, 70, and 100%.

3 Methods

3.1 Culture of Bacterial Colonies

1. Prepare 100 mL N.B. media by dissolving 13 gm of media in purified water.
2. Autoclave prepared media along with other materials: 15 and 50 mL centrifuge tubes, micro-centrifuge tubes, micro tips, spreader.
3. Prepare agar plate.
4. Pipette 20 μL bacterial colonies from the stock (20 μL is sufficient) and streak on the agar plate in quadrant manner. Incubate agar plate at 37 °C overnight (*see* **Note 1**).
5. After sufficient colony growth, transfer one colony to the 50 mL of autoclaved broth and incubate at 37 °C.
6. After getting sufficient growth, 10 μL of the culture from **step 6** is transferred to the fresh broth (*see* **Note 2**).
7. Adjust the optical density of the colonies at 0.1 a.u. by recording absorption at 600 nm.
8. This is equivalent to the 10^8 colony-forming unit/mL (CFU/mL).

3.2 Preparation of Peptide Hydrogels and Antimicrobial Assay

1. Prepare the hydrogel in a 96 well plate in triplicates and seal with parafilm.
2. Incubate the prepared hydrogel at 37 °C for 2 h.

3. After 2 h, add nutrient broth (N.B.) media (or Luria broth media as per the bacterial species) to the hydrogel wells for equilibration and incubate at 37 °C for 12 h (overnight).
4. After 12 h (before the assay) remove N.B. media and replace with 200 μL fresh media.
5. Discard the 200 μL fresh media and add 100 μL of bacterial colonies from freshly cultured bacterial colonies (Subheading 3.1 **step** 7).
6. Prepare serial dilution of bacterial colonies (Subheading 3.1 **step** 7) and add to the prepared hydrogels in 96 well plate in triplicates.
7. Add bacterial colonies in 96 well plate in triplicate, adding peptide hydrogel.
8. Prepare peptide hydrogel in 96 wells for negative control (without bacterial colonies).
9. Incubate the 96 well plates at 37 °C for 24 h.
10. After 24 h, transfer the bacterial colonies from the top of the hydrogel to the cuvette, and record optical density at 600 nm.

3.3 Outer Membrane PermeabilizationAssay

1. Prepare 0.5 mM stock solution of 1-N-phenylnaphthylamine (NPN) in acetone (*see* **Note 3**).
2. Centrifuge the bacterial colonies from secondary culture at 3500× *g* for 5 min, 4 °C, and wash with PB (7.4 pH) twice.
3. Dilute the washed bacterial colonies to 10^8 cells and transfer 100 μL of washed bacterial cells to the peptide hydrogel prepared in 96 well plate as explained, and incubate for 2 h.
4. After 2 h, transfer the bacterial cells to the cuvette and add NPN dye (final concentration, 10 μM).
5. Incubate polymyxin B (10 μg/mL) with bacterial cells (positive control) [9] as described in **step 3** and proceed with **step 4**.
6. Set the spectrofluorometer instrument at excitation wavelength 356 nm and emission wavelength 425 nm.
7. Record the fluorescence emission intensity for bacterial cells and polymyxin B.

3.4 Determination of Intracellular Reactive Oxygen Species (ROS) Generation

1. Prepare 2,7-dichlorofluorescein diacetate (DCFH2-DA) solution in ethanol (10 mM stock solution).
2. Culture the bacterial cells as explained.
3. Prepare peptide hydrogels in 96 well plate.
4. Incubate 200 μL bacterial colonies with peptide hydrogels and incubate for 12 h.
5. Add 5 μL DCFH-2-DA at 37 °C for 3 h (10 μM final concentration).

6. Set the spectrofluorometer at the excitation wavelength at 503 nm and emission wavelength at 523 nm.
7. Record the fluorescence emission intensity.

3.5 FESEM Analysis (Membrane Disruption Assay)

1. Centrifuge the bacterial culture of interest to get the cell pellet at 5000× *g* for 5 min.
2. Wash the cell pellets with 1× phosphate buffer at room temperature and again centrifuge at 3000× *g* for 5 min to get the final cell pellet (*see* **Note 4**).
3. Add primary fixative solution 2.5% glutaraldehyde to the samples and incubate at 4 °C for 3 h.
4. Wash twice with phosphate buffer and let the cells air dry, followed by washing with deionized water and alcohol gradient ranging from 30 to 100%, increasing with 10% alcohol in each wash and then allowing the cells air dry.
5. Coat the cells with gold and image using FESEM.

4 Notes

1. *To prevent contamination of bacterial culture:* Perform Gram staining to ensure the contamination. Wipe the working place properly with 70% ethanol or 70% isopropanol and sterilize the laminar flow cabinet with UV-C lamp for 20 min. Turn on the spirit burner and sterilize the conical flask opening before and after every pipetting. Try to maintain minimum movement and avoid touching inside the laminar flow cabinet. Do not overcrowd the laminar flow cabinet, because it may block the proper airflow [10].
2. *If there is no growth in bacterial culture:* Check your bacterial stock solution, whether it is properly prepared and stored at −20 °C (do not use same vial for more than three times). Plot a growth curve and check the doubling time of the bacterial colonies. Check the incubation temperature [10].
3. *Problem in ROS and outer membrane permeabilization assays:* Prepare fresh solution (dilution) and store the stock solution in −20 °C (can be used up to 3 months). Check the pH of buffer. Optimize the incubation temperature and time [7].
4. *FESEM imaging:* Wash bacterial cells properly and do not increase the centrifuge speed (rpm) abruptly. Optimize the incubation temperature and time after adding 2.5% glutaraldehyde (4 °C and 3 h is the standard incubation temperature and time). Sample should be completely dry before sputtering. Optimize the coating step (sputtering), whether single or double coating is required. Optimize the accelerating voltage [11].

References

1. Goyal R, Jerath G, Chandrasekharan A et al (2020) Peptide-based delivery vectors with pre-defined geometrical locks. RSC Med Chem 11:1303–1313
2. Prakash V, Mukesh B, Sasidharan S et al (2022) Syndiotactic hexamer peptide nanodots. Eur Biophys J 51:483–491
3. Das AK, Gavel PK (2020) Low molecular weight self-assembling peptide-based materials for cell culture, antimicrobial, anti-inflammatory, wound healing, anticancer, drug delivery, bioimaging and 3D bioprinting applications. Soft Matter 16:10065–10095
4. Goyal R, Jerath G, Akhil R et al (2021) Geometry encoded functional programming of tumor homing peptides for targeted drug delivery. J Control Release 333:16–27
5. Hazam PK, Akhil R, Jerath G et al (2019) Topological effects on the designability and bactericidal potency of antimicrobial peptides. Biophys Chem 248:1–8
6. Salick DA, Kretsinger JK, Pochan DJ et al (2007) Inherent antibacterial activity of a peptide-based β-hairpin hydrogel. J Am Chem Soc 129:14793–14799
7. Prakash V, Christian Y, Redkar AS et al (2022) Antibacterial hydrogels of aromatic tripeptides. Soft Matter 18(34):6360–6371
8. Gallagher AG, Alorabi JA, Wellings DA et al (2016) A novel peptide hydrogel for an antimicrobial bandage contact lens. Adv Healthc Mater 5:2013–2018
9. Saikia K, Chaudhary N (2018) Antimicrobial peptides from C-terminal amphipathic region of E. coli FtsA. Biochim Biophys Acta Biomembr 1860:2506–2514
10. Nambisan P (2017) An introduction to ethical, safety and intellectual property rights issues in biotechnology. Academic Press
11. Saikia K, Sravani YD, Ramakrishnan V, Chaudhary N (2017) Highly potent antimicrobial peptides from N-terminal membrane-binding region of E. coli MreB. Sci Rep 7(1):1–9

Chapter 17

Rheology of Peptide Based Hydrogels

Kalpana Kumari and Vibin Ramakrishnan

Abstract

Hydrogels have been established as useful biomaterials for the development of drug-delivery vehicles. Hydrogels are preferred for such applications due to their similarity to the host's extracellular matrix. It has high water content, relatively high biocompatibility, and self-healing properties. It can also be used as a heterogeneous catalyst for industrial applications. Hence, it is crucial to determine the rheological properties of hydrogels to make them more efficient in their application. This chapter aims to outline the experimental protocols that can ascertain some important rheological properties of peptide hydrogels.

Key words Hydrophilic, Self-healing, Visco-elastic, Hydrogels, Rheometer, Sweep test, HFIP

1 Introduction

Peptide hydrogels are extensively used in cell culture and tissue engineering. Hydrogels are hydrophilic 3D solid-like networks and have visco-elastic properties. They closely resemble the extracellular matrix and several soft biological tissues [1–4]. Hydrogels have high water content and thus can be used for various biomedical applications. They are classified into different categories on the basis of their mechanical properties, mode of preparation, and extent of biodegradation. Hence, Rheometer evolved as an important tool to determine the mechanical and rheological features of the hydrogels at the micro and nano level. This instrument is fast, sensitive, and uses a small amount of sample [5–8]. The rheological test of hydrogels based on various parameters is presented in this protocol. The self-healing feature of peptide hydrogels has important applications in the development of drug-delivery systems and tissue engineering. Self-healing peptide hydrogels are modified into low-viscous liquid in the presence of high strain, and the structure is retained upon removal of strain [9]. Due to its mechanical feature, it also provides support for cell attachment. The amplitude sweep test is used to study physical strength of peptide hydrogels. Storage

Vibin Ramakrishnan (ed.), *Biophysical Characterization of Functional Peptides*, Springer Protocols Handbooks,
https://doi.org/10.1007/978-1-0716-3405-9_17,

modulus, loss modulus, deforming strain, and linear visco-elastic region can be studied. The frequency sweep test is also used to further validate the strength of the peptide hydrogel. If solid-state component (G') is higher than liquid-state component (G"), it represents high strength and gel-like behavior. This study outlines a working protocol to assess rheological properties such as amplitude sweep test, frequency sweep test, and self-healing property of peptide hydrogels.

2 Materials

1. Rheometer. Principal suppliers: AMETEK Brookfield, Anton Paar, Alpha technologies, NETZSCH, SUCEEDER.
2. Peptide hydrogel (1–2%).
3. Methanol.
4. HFIP.
5. Lint free tissue.

3 Methods

3.1 Instrument Preparation

1. Switch on the pump before operating the instrument.
2. Fix the parallel plate disc on the holder before adding sample over the platform of rheometer. Sample is added between two shearing plates. The gap between the disks is usually <1 mm (*See* **Note 2**).

3.2 Sample Preparation

Preparation of 100 mM phosphate buffer: Weigh 600 mg of monosodium phosphate (NaH_2PO_4) and dissolve it in 50 mL of deionized water. Check the pH before using it to prepare working solution. Prepare 25 mM phosphate buffer for the preparation of hydrogel.

Preparation of 2% hydrogel: Weigh 10 mg of the purified and lyophilized peptide and dissolve in the 50 μL HFIP (≤ 10%) of total volume if it is insoluble in buffer. Add 450 μL of 25 mM phosphate buffer (pH 7.4) to make the final volume to 500 μL. Vortex it. Incubate for 5–10 min. It usually results in formation of gels (*See* **Note 1**).

3.3 Data Collection

1. Once the sample is placed on the platform and keeping zero gap between the plates, select the type of test that needs to be performed for the hydrogel (*See* **Note 2**). For amplitude sweep, perform the experiment at constant frequency (1 Hz).
2. For frequency sweep test, experiment may perform in the range of angular frequency from 0.1 to 100 s^{-1} at constant strain (under the linear visco-elastic region).

3. For the determination of self-healing property of hydrogel, step-strain dependent experiment needs to be performed. Strain will be varied from low shear strain of 0.1–300% and followed by recovering at low shear strain. Repeat the step-strain cycle thrice.

3.4 Data Analysis

1. The amplitude sweep test is used to determine storage modulus (G') and loss modulus (G") with deforming strain (%). This test is also used to determine linear visco-elastic regions of hydrogels. Hydrogels are returned to its original structure once the strain is removed within the LVE region.
2. The frequency sweep test is performed to further validate the physical strength and cohesiveness of hydrogels. It is used to determine G' and G" as a function of frequency at constant amplitude. It measures the effect of oscillatory frequency on the dynamic storage and loss modulus at fixed temperature (*See* **Note 3**).
3. For self-healing rheological test, hydrogels are subjected to high deforming shear and then allowing it to recover at low shear [9].

4 Notes

1. Hydrogels should be freshly prepared.
2. Before taking the readings, trim the extra sample of hydrogel after placing it between the plates.
3. Frequency sweep analysis must be performed within the (linear visco-elastic) LVE region [9].

References

1. Zuidema JM, Rivet CJ, Gilbert RJ et al (2014) A protocol for rheological characterization of hydrogels for tissue engineering strategies. J Biomed Mater Res B Appl Biomater 102: 1063–1073
2. Fuentes-Caparras AM, Canales-Galarza Z, Barrow M et al (2021) Mechanical characterization of multilayered hydrogels: a rheological study for 3D-printed systems. Biomacromolecules 22:1625–1638
3. Rammensee S, Huemmerich D, Hermanson KD et al (2006) Rheological characterization of hydrogels formed by recombinantly produced spider silk. Appl Phys 82:261–264
4. Terri-Chen T (2017) Using rheology to characterize flow and viscoelastic properties of hydrogels, Adhesives and Biopolymers 1–58
5. Stojkov G, Niyazov Z, Picchioni F et al (2021) Relationship between structure and rheology of hydrogels for various applications. Gels 7:255
6. Alam K, Iqbal M, Hasan A et al (2020) Rheological characterization of biological hydrogels in aqueous state. J Appl Biotechnol Rep 7:172–176
7. Saikia J, Dharmalingam K, Anandalakshmi R et al (2021) Electric field modulated peptide based hydrogel nanocatalysts. Soft Matter 17: 9725–9735
8. Rodríguez-Llansola F, Miravet JF, Escuder B (2009) A supramolecular hydrogel as a reusable heterogeneous catalyst for the direct aldol reaction. Chem Commun (Camb) 47:7303–7305
9. Prakash V, Christian Y, Redkar AS et al (2022) Antibacterial hydrogels of aromatic tripeptides. Soft Matter 18:6360–6371

Chapter 18

Estimation of Aggregation Kinetics

Jahnu Saikia and Vibin Ramakrishnan

Abstract

Amyloid formation is central to the progressive pathology of several human diseases, such as Alzheimer's, Huntington's, type II diabetes, and spongiform encephalopathy. The formation of amyloid fibrils from disordered peptides occurs through a nuclei dependent polymerization pathway characterized by the slow nucleation phase followed by the rapid oligomerization and fibril formation; also called the elongation phase. More recently, they have also found application in the fabrication of various functional materials. It is important to generate monomeric units and understand the factors governing their kinetics for both application fronts: therapeutic intervention and novel materials. In this chapter, we have considered $A\beta_{1-42}$ as a model peptide and discussed a reliable protocol to produce monomeric units which can be used as a starting material for kinetic experiments. Further, this chapter details different biophysical techniques such as fluorescence spectroscopy, circular dichroism, Fourier transform infrared spectroscopy, dynamic, static light scattering and transmission electron microscopy to characterize $A\beta_{1-42}$ nucleation, aggregation, and fibrillization.

Key words Amyloid fibrils, Fibrillar morphology, Thioflavin T fluorescence assay, Determination of aggregation formation, Transmission electron microscopy (TEM), Fourier-transform infrared spectroscopy (FTIR), Circular dichroism (CD), Dynamic light scattering (DLS)

1 Introduction

Amyloids are protein aggregates often characterized by largely similar fibrillar morphology and physicochemical features. The dysfunctional assembly of proteins and peptides results in various amyloid pathologies such as Alzheimer's disease, type II diabetes, Parkinson's disease, and prion diseases [1, 2]. Over the years, researchers have realized that amyloid aggregates are not only related to diseased states but can be exploited to form functional biomaterials. While designing synthetic peptide materials, protein aggregation may be viewed as a "feature" rather than a "bug." This recognition has inspired extensive efforts to develop novel peptide-based biomaterials and has found applications in biomedicine, energy storage, and catalysis [3–5].

Vibin Ramakrishnan (ed.), *Biophysical Characterization of Functional Peptides*, Springer Protocols Handbooks,
https://doi.org/10.1007/978-1-0716-3405-9_18,

Alzheimer's disease (AD) is the most frequent type of human amyloidosis and constitutes approximately 80% of all dementia cases worldwide. Brain pathology of an AD patient suggests extra-cellular and intra-neuronal neurofibrillary tangles of Aβ in the form of amyloid plaques and cerebral amyloid angiopathy that affects the cognitive areas of the brain. According to the "Aβ hypothesis," the aggregation of Aβ proceeds through the rate-limiting nucleation-dependent fibrillation phase, followed by an exponential elongation phase to form full-length fibrils. The ability to produce monomeric or oligomeric Aβ of smaller sizes (such as dimers, trimers, and tetramers) and the investigation of the mechanistic function these particles serve in uncovering potential therapeutic approaches for AD.

In this chapter, we describe a reliable protocol to produce monomeric $A\beta_{1-42}$, which will be used as a starting material for kinetic experiments. Further, this chapter details different biophysical techniques such as fluorescence spectroscopy, circular dichroism, Fourier transformed infrared spectroscopy, dynamic and static light scattering, and transmission electron microscopy to characterize aggregation and fibrillization in a global manner.

2 Materials

Preparation of LMW Aβ aggregates for determination of aggregation kinetics: Trifluoroacetic acid/1,1,1,3,3,3-hexafluoro-2-propanol (TFA/HFIP), Aβ peptide, vacuum desiccator, spectrofluorometer, DMSO, phosphate buffer (50 mM, pH 7.4), thioflavin T, quartz cuvette with 1 cm path length, nitrogen gas for purging.

3 Methods

3.1 Preparation of LMW Aβ Aggregates Using 1,1,1,3,3,3-Hexafluoro-2-Propanol (HFIP) Pre-treatment

TFA/HFIP is used to induce alpha-helical conformations. This is intended to produce monomeric solutions and abolish the peptides' conformational memory. It is followed by the removal of TFA/HFIP and re-suspension using an aqueous media, or any compatible re-suspension solvent such as DMSO/phosphate buffer.

1. Dissolve $A\beta_{1-42}$ (0.5 mg) in 1 mL of TFA. This will generally result in a homogeneous solution. Aliquots of 30 μL in 0.5 mL Eppendorf tubes were used.
2. TFA was evaporated using N_2 gas. HFIP is used to remove TFA that is left behind after **step 1** followed by evaporation using a vacuum desiccator.
3. The peptide was dissolved in 50 mM phosphate buffer at pH 7.4.

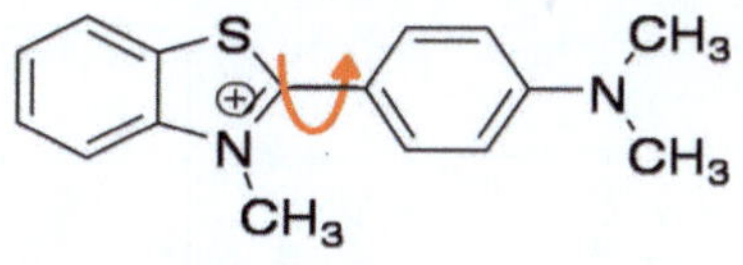

Fig. 1 Structure of ThT

4. To verify the presence (or absence) of any pre-aggregates in the peptide stock solution after treatment with TFA or HFIP, 90° static light scattering experiment has been performed. Parameter settings: λ_{Ex}:450 nm, slit width = 2.5 nm; λEm:450 nm, slit width = 2.5 nm).

3.1.1 Stock Preparation

1. Add 40.5 μL of dimethyl sulfoxide (DMSO) to the dried peptide to get a peptide stock of 50 mM concentration (vortex to dissolve).
2. Add 9 μL peptide stock (50 mM, 100% DMSO) to the MCT and incubate for 5 h.

3.2 Thioflavin T Fluorescence Assay

In 1959, Vassar and Culling reported the use of the benzothiazole dye thioflavin T (ThT). ThT is a potent fluorescent marker of amyloid (Fig. 1). They observed that ThT gets selectively localized to amyloids, thus showing an increase in fluorescence. ThT linearly correlates with the concentration of amyloid over ThT concentrations at a range from 0.2 to 500 μM [6].

Naiki et al. and Levine reported that upon binding with fibrils, ThT shows significant increase in absorbance at 482 nm and that ThT fluorescence results only when the dye gets bound to amyloid fibrils [7, 8]. The rotational immobilization of the C–C bond, connecting the aniline and benzothiazole rings upon binding to amyloid fibrils, results in fluorescence enhancement [9, 10]. ThT binds to the side chain channels along the long axis in a span of four consecutive β-strands. This makes ThT a very sensitive and efficient reporter of diverse fibrils prepared of biological origin in experimental systems (K_d in the sub and low μM range) [11].

3.2.1 ThT Fluorescence Assay Time Scan

1. ThT binding experiments are performed by combining 20 μL of 3 mM peptide to a 580 μl solution of 10 μM ThT in 50 mM sodium phosphate buffer having 6% DMSO, pH 7.5.
2. Fluorescence measurements are recorded in a Horiba Florimax-4 spectrofluorometer at 25 ° C using a 1 cm quartz cuvette. The excitation wavelength was set at 450 nm (slit width = 2.5 nm), and fluorescence emission is monitored at 485 nm (slit width = 5 nm) for 12 min. Quantification of fibrillar aggregates formed in the solution can be done by measuring the relative fluorescence at 485 nm.
3. ThT fluorescence is recorded in the following order at 0th, 6th, 12th, and 24th hour:

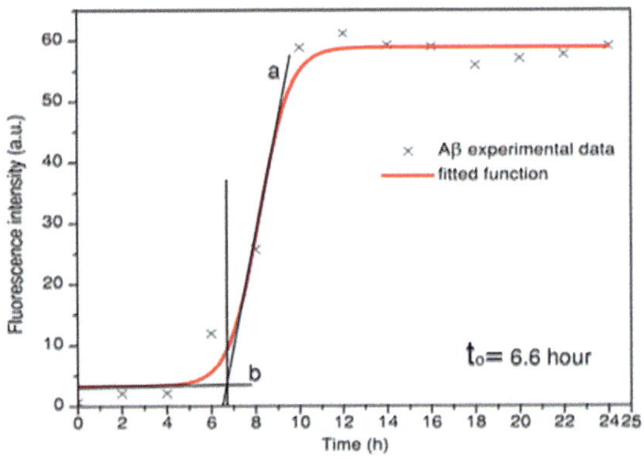

Fig. 2 Aggregation kinetics of $A\beta_{1-42}$ peptide. ThT fluorescence assay is measured as a function of time at 100 μM concentration. During the time course of aggregation, $A\beta_{1-42}$ follows a sigmoidal curve. It has a typical lag phase followed by a steep exponential phase followed by a plateau. (Reprinted with permission from ACS Chem. Neurosci. 2019, 10, 5, 2250–2262. Copyright 2019 American Chemical Society)

(a) Buffer (150 μL, 200 mM) + DMSO (36 μL) + deionized water (414 μL)

(b) (1) + 6 μL ThT

(c) Peptide 100 μM (20 μl 3 mM peptide +150 μL (200 mM) buffer, 34.8 μL DMSO +395.2 μL deionized water)

(d) (3) + 6 μL ThT

3.2.2 Determination of Lag Phase during Aggregation

1. Dissolve 100 μM $A\beta_{1-42}$ peptide monomers in 50 mM phosphate buffer (pH 7.4).
2. Add 10 μM ThT to the solution before measuring ThT fluorescence at 25 ° C.
3. ThT fluorescence was measured as a function of time for the next 24 h at a regular time interval (Fig. 2).

4 Morphological Characterization of $A\beta_{1-42}$ Aggregates

Transmission electron microscopic images help to examine the morphological appearance of $A\beta_{1-42}$ amyloid aggregates in water.

4.1 Transmission Electron Microscopy (TEM) Analysis

1. $A\beta_{1-42}$ (20 μL) was placed on carbon-coated copper mesh grids, and the residual buffer was carefully wicked away.
2. The samples are then negatively stained using 2% (w/v) saturated uranyl acetate.
3. The excess stain was wicked away after 30 s. The grids were allowed to air dry.

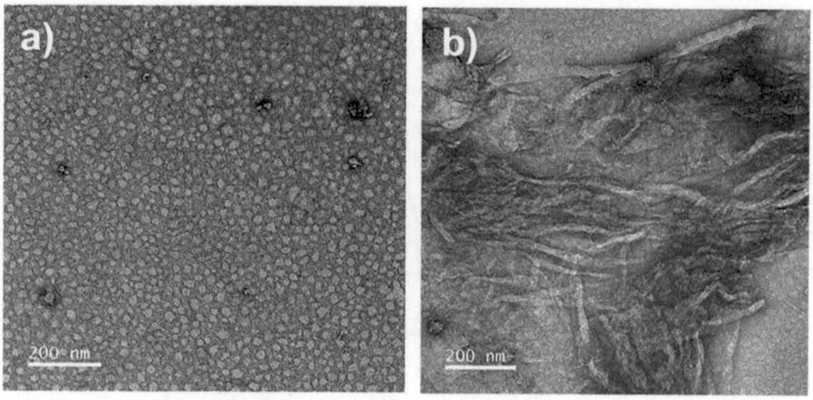

Fig. 3 TEM of Aβ1–42 after 0th and 24th h of incubation under ambient conditions. The sample after 24 h has fibrils as the predominant form developed from the initial homogenous spherical oligomeric clusters. Scale bars correspond to 200 nm. (Reprinted with permission from ACS Chem. Neurosci. 2019, 10, 5, 2250–2262. Copyright 2019 American Chemical Society)

4. JEM-2100F field emission transmission electron microscope (JEOL, Tokyo, Japan) at 200 kV was used for imaging (Fig. 3).

4.2 Secondary Structure Analysis of the Peptide Aggregates: Both FTIR and CD Are Employed for the Analysis of the Change in the Secondary Structure

4.2.1 FTIR Spectroscopic Measurements

1. $A\beta_{1-42}$ is cast on a potassium bromide (KBr) pellet, re-suspended with D_2O, and vacuum dried for dry measurements.
2. Scanned between 1400 cm^{-1} to 1800 cm^{-1} (amide 1). Micro FTIR-200 (Jasco Co., Japan) was used. It is equipped with an MCT detector at 4 cm^{-1}.

4.2.2 Circular Dichroism Measurements

1. Circular dichroism (CD) spectra were recorded in a quartz cell with a slit width of 1 nm and pathlength 1 mm, using a Jasco J-1500 spectropolarimeter.
2. Peptides were dissolved in 10 mM phosphate buffer and made up to final concentration of 10 μM.
3. Spectra were recorded from 260 to 190 nm at room temperature.
4. The spectra are presented as mean residue ellipticity, $[\theta]_{MRE}$.
5. Mean residue ellipticity was calculated using the following relation:

   $[\theta]_{MRE} = (M_r \times \theta_{mdeg})/(10 \times l \times c)$, where M_r is the mean residue weight. θ_{mdeg} is the ellipticity in milli-degrees, l is the path length in centimeters, and *c* is the peptide concentration in mg mL^{-1}.

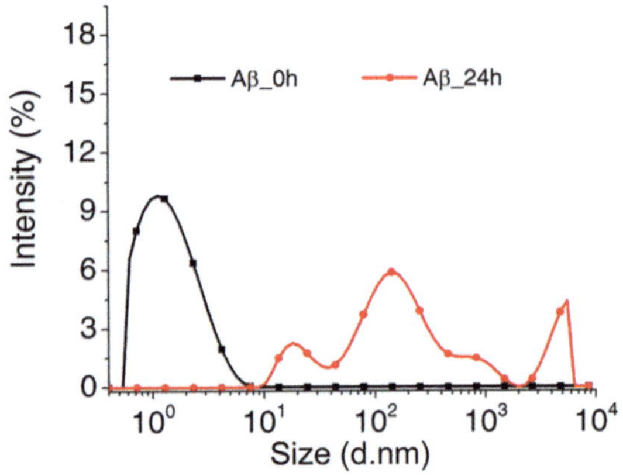

Fig. 4 Changes in the hydrodynamic radius of $A\beta_{1-42}$ aggregates monitored by DLS as a function of time. Samples containing 100 μM peptide were added to a quartz cuvette with a 1 cm path length. The size distribution of the aggregates was monitored at two time points, 0th and 24th hour. (Adapted with permission from ACS Chem. Neurosci. 2019, 10, 5, 2250–2262. Copyright 2019 American Chemical Society)

4.3 Dynamic Light Scattering (DLS)

DLS experiments were performed on a Zetasizer Nano ZS. Dynamic light scattering is the most widely employed technique for size and polydispersity measurements in protein, polymer, and nanoparticle research.

1. A laser source with $\lambda = 633$ nm and a detector at a scattering angle of $\theta = 173$ degrees is used.
2. Samples with 100 μM (100 μL) $A\beta_{1-42}$ after 24 h of incubation were analyzed and compared with freshly prepared samples (control).
3. All measurements are repeated ten times and mean value is considered for analysis.
4. Zetasizer program was used to measure the average diameter (Fig. 4).

References

1. Cohen FE, Kelly JW (2003) Therapeutic approaches to protein-misfolding diseases. Nature 426:905–909
2. Saikia J, Pandey G, Sasidharan S, Antony F, Nemade HB, Kumar S, Chaudhary N, Ramakrishnan V (2019) Electric field disruption of amyloid aggregation: potential noninvasive therapy for Alzheimer's disease. ACS Chem Neurosci 10:2250–2262
3. Saikia J, Dharmalingam K, Anandalakshmi R, Redkar AS, Bhat VT, Ramakrishnan V (2021) Electric field modulated peptide based hydrogel nanocatalysts. Soft Matter 17:9725–9735
4. Saikia J, Bhat VT, Potnuru LR, Redkar AS, Agarwal V, Ramakrishnan V (2022) Minimalist de novo design of an artificial enzyme. ACS Omega 7(23):19131–19140

5. Pandey G, Saikia J, Sasidharan S, Joshi DC, Thota S, Nemade HB, Chaudhary N, Ramakrishnan V (2017) Modulation of peptide based nano-assemblies with electric and magnetic fields. Sci Rep 7:1–9
6. CF, C. (1959) Fluorescent stains, with special reference to amyloid and connective tissues. Arch Pathol 68:487–498
7. Naiki H, Higuchi K, Hosokawa M, Takeda T (1989) Fluorometric determination of amyloid fibrils in vitro using the fluorescent dye, thioflavine T. Anal Biochem 177:244–249
8. Levine H III (1993) Thioflavine T interaction with synthetic Alzheimer's disease β-amyloid peptides: detection of amyloid aggregation in solution. Protein Sci 2:404–410
9. Virchow R (1853) Nouvelles observations sur la substance animale analogue à la cellulose végétale. CR Acad Sci (Paris) 37:860–861
10. Howie AJ, Brewer DB, Howell D, Jones AP (2008) Physical basis of colours seen in Congo red-stained amyloid in polarised light. Lab Investig 88:232–242
11. Groenning M (2010) Binding mode of thioflavin T and other molecular probes in the context of amyloid fibrils—current status. J Chem Biol 3:1–18

Chapter 19

Assessment of Amyloid Morphology Using Electron Microscopy

Laxmi Raman Adil and Vibin Ramakrishnan

Abstract

Aβ amyloid fibrils have been demonstrated to be vital for triggering Alzheimer's disease, as reported by numerous research works. Therefore, Aβ amyloid fibrils formation is considered a hallmark for detecting Alzheimer's disease (AD). Anomalous multiplications of the Aβ precursor protein (APP) gene and mutation of the APP gene lead to AD. Understanding the amyloid formation mechanism could help us design new technologies and methods for its prevention and treatment. Hence it is vital to comprehend the amyloid structure and self-assembly morphology. Many research techniques are available to study peptides, oligomers, and fibrils. Verities of the experimental methods for visualizing the amyloid morphology are still expanding. The electronic microscopy technique is the most widely used approach for assessing amyloid morphology.

Key words Aβ amyloid, Alzheimer's disease (AD), Self-assembly morphology, Field-emission transmission electron microscope (FETEM), Field-emission scanning electron microscope (FESEM)

1 Introduction

Amyloid aggregates are associated with human disease, which afflicts millions of lives globally [1]. Amyloid formed by protein and peptides has a tendency to form aggregates. Amyloid refers to abnormal, extracellular, proteinaceous fibrous deposits. Amyloid tends to form aggregates and has a dominant β-sheet structure [2, 3]. Generally, normal fiber protein has motility, supportive, and structural roles, but amyloid is associated with various diseases, including type 2 diabetes mellitus (T2DM) and Alzheimer's disease [4, 5]. Generally, the standard microscope uses lightbulbs for illumination of the samples having a resolution of up to 1500X. For the source of illumination, the electron used in electron microscopy resulted in higher resolution up to 300,000X [6]. Electron microscopy (EM) is a technique for obtaining high-resolution images of biological and non-biological specimens. It is used in biomedical

Vibin Ramakrishnan (ed.), *Biophysical Characterization of Functional Peptides*, Springer Protocols Handbooks,
https://doi.org/10.1007/978-1-0716-3405-9_19,

research to investigate the detailed structure of tissues, cells, organelles, and macromolecular complexes. There are two main types of EM generally used, they are as follows.

1.1 Field-Emission Transmission Electron Microscope (FETEM)

In the field-emission transmission electron microscope (FETEM) technique electron beam is transmitted through a sample that is further translated to form an image. Furthermore, the image of the sample is recorded by the focus of the translated beam onto an imaging device, such as a photographic film or fluorescent screen.

1.2 Field-Emission Scanning Electron Microscope (FESEM)

It is based on a raster scanning technique to visualize and magnify the images of the specimen. A beam of the electron is directed toward the specimen, in which beam loses its energy via passes through. The losing energy explains/translates the information about specimen composition, topography, and morphology. We can visualize small particles and the morphology of different types of samples by using the FESEM instrument.

2 Materials

2.1 Required Materials for Preparing a Sample for FETEM

1. Amyloid β (Aβ) stock solution.
2. Multichannel pipette set and tips.
3. Ultrapure water.
4. Lint-free tissue paper.
5. Carbon-coated copper mesh TEM grids (300 mess).
6. Uranyl acetate solution (2% w/v).
7. Petri dish.
8. Stainless steel tweezer.

2.2 Required Materials for Preparing a Sample for FESEM

1. Amyloid β (Aβ) stock solution.
2. Multichannel pipette set and tips.
3. Ultrapure water.
4. Lint-free tissue paper.
5. Petri dish.
6. Stainless steel tweezer.
7. Glass slide.
8. Carbon tape.

3 Method

3.1 Sample Preparation Method for FETEM

1. Wash copper grids using distilled or ultrapure water and pull out excess water by lint-free tissue paper.
2. Drop-cast amyloid β (4 μL of 99 μM) sample for 2 min on a copper grid (*see* **Note 1** and **Note 2**).
3. Use 2% (w/v) uranyl acetate solution on the TEM grid for negatively staining.
4. Thirty seconds later, remove the extra liquid by using lint-free tissue paper from the TEM grid (*see* **Note 3**).
5. In the next step for the sample examination and image collection from the TEM grid use a 200-kV operated electron microscope and save it to the respective folder.

3.2 Instrument Preparation for FETEM

1. Fill the liquid nitrogen for the creation of a vacuum in the instrument.
2. Check the vacuum in the switch selector it should be in the 10^{-5} Pa range.
3. Open the software "TEM Center" on the desktop.
4. In the high voltage control window please make sure the voltage is on (200 kV).
5. Gently remove the specimen rod from the goniometer by rotating anticlockwise and place it on the holder stand.
6. Samples have been loaded into the specimen blade with clean tweezers and tighten the screw very gently.
7. In the last insert the specimen rod in the sample holder, analyze and save TEM images for further application (Fig. 1).

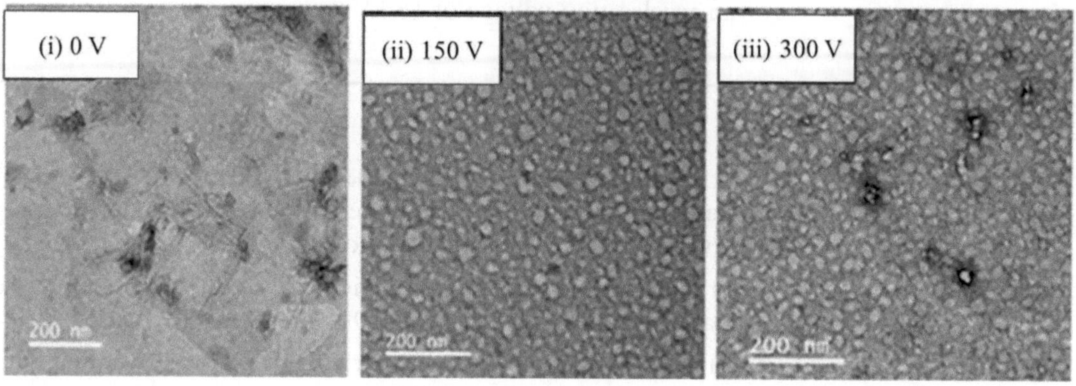

Fig. 1 FETEM image of A$\beta_{1\text{-}42}$ (**a**) after incubation for 24 h in normal conditions, the predominant form of the standard sample has fibrils, whereas (**b**) and (**c**) with the external electric field where sample kept in 150 V cm^{-1} and 300 V cm^{-1} their spherical oligomer clusters observed (Scale 200 nm). (Reprinted with permission from ACS Chem. Neurosci. 2019, 10, 5, 2250-2262. Copyright 2019 American Chemical Society)

3.3 Instrument Preparation for FESEM

1. After switching on the instrument, open the software for "SmartSEM" software in the integrated desktop.
2. Please confirm that the sample stage is in the initial stage and all the other parameters like gun vacuum, system vacuum, etc. are showing correct values.
3. In the next step switch off the EHT.
4. Close the gun vacuum column chamber and vent the instrument sample chamber by filling inert gas.
5. Place the biological samples in sample stabs, load it inside the sample chamber and analyze properly and save the amyloid pictures for further studies.

3.4 Sample Preparation Method for FESEM

The surface of the biological samples should be conductive for the FESEM imaging experiment. But generally, most of the biological specimens are non-conductive in nature. Therefore, biological specimens cannot be visualized in the native state and need to the coating of a conductive layer on the surface of the specimen (*see* **Note 4**). Figure 2 shows morphology analysis of the Aβ fibrils in different condition demonstrated [7]. The method for sample preparation for FESEM is as follows:

1. Drop caste 2 μL of specimen sample on the surface of the glass slide/silicone sheet.
2. Keep the sample in a dust-free environment overnight to dry.
3. Mount the sample on the metal stub using a sticky carbon tape.
4. To make the surface of the specimen sample conductive, it has been coated with gold or platinum by a high-resolution sputter (JEOL JFC-1600) coater at 30 mA for 2 s.
5. The specimen was examined by the JEOL JSM 7400F system at an acceleration voltage of 15 kV.

4 Notes

1. For FETEM and FESEM experiments amyloid (biological) sample preparation should be performed very carefully by preserving its three-dimensional morphology.
2. Prepared specimen must be dried, cleaned, and it should be stable in the higher vacuum (10^{-5} pa).
3. Amyloid sample must be properly fixed on the substrate.
4. Coating gold is a critical step, it must be thick sufficient to prevent charging and prevent from burning, whereas it should not be thicker than it hinders sample surface details.

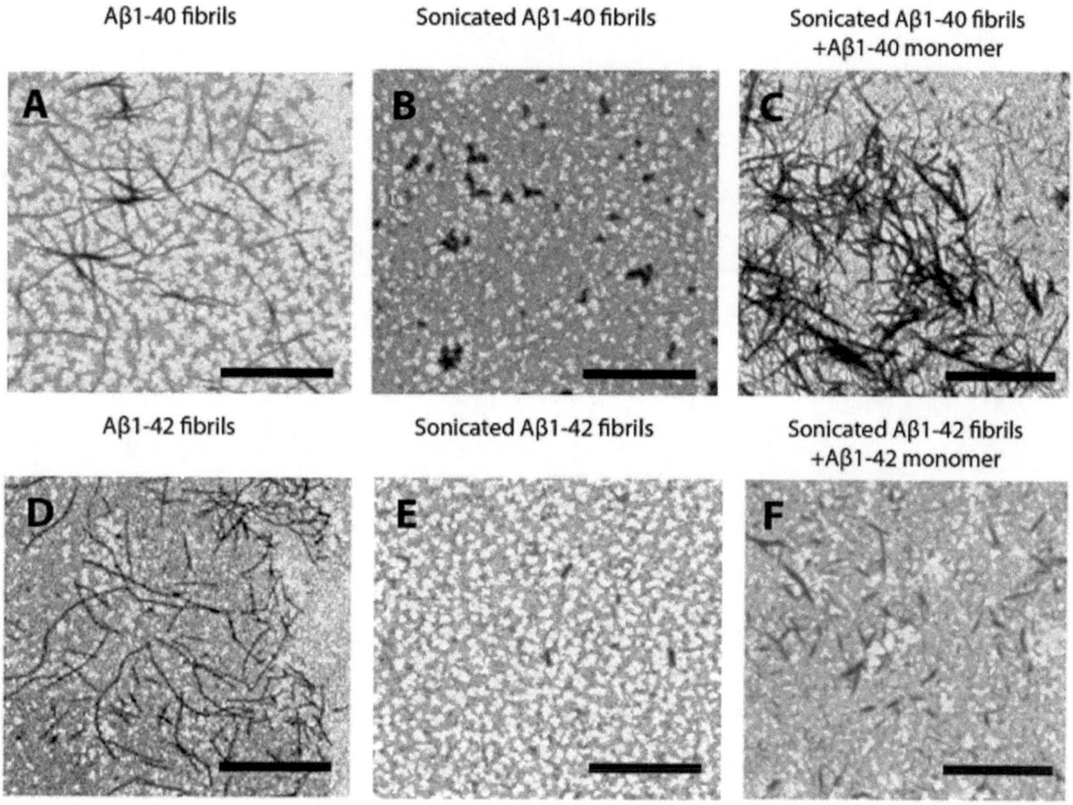

Fig. 2 FESEM morphology analysis of the Aβ fibrils. (**a**) Under stagnant incubation conditions A$\beta_{1\text{-}40}$ fibrils, (**b**) microtip sonication of A$\beta_{1\text{-}40}$ fibrils to obtain very short fibrillar pieces, (**c**) monomeric Aβ_{1-40} with sonicated and immobilized A$\beta_{1\text{-}40}$ fibrils probed, (**d**) fibrils of A$\beta_{1\text{-}42}$ prepared stagnant incubation conditions followed by immobilization on the surface of SPR chip, (**e**) fibrils of A$\beta_{1\text{-}42}$ microtip sonication to get very short fibrillar pieces, and immobilized on the surface of chip, (**f**) the immobilized and sonicated fibril of Aβ_{1-42} probed with monomeric Aβ_{1-42}. (Reproduced with permission from Data in Brief, 19, 2018, 1166-1170. Copyright 2018 The Authors. Published by Elsevier Inc.)

References

1. Saikia J, Pandey G, Sasidharan S, Antony F, Nemade HB, Kumar S, Chaudhary N, Ramakrishnan V (2019) Electric field disruption of amyloid aggregation: potential noninvasive therapy for Alzheimer's disease. ACS Chem Neurosci 10:2250–2262
2. Benilova I, Karran E, Strooper BD (2012) The toxic Aβ oligomer and Alzheimer's disease: an emperor in need of clothes. Nat Neurosci 15: 349–357
3. Blanchard JW, Victor MB, Tsai L-H (2022) Dissecting the complexities of Alzheimer disease with in vitro models of the human brain. Nat Rev Neurol 18:25–39
4. Economou NJ, Giammona MJ, Do TD, Zheng X, Teplow DB, Buratto SK, Bowers MT (2016) Amyloid β-protein assembly and Alzheimer's disease: dodecamers of Aβ42, but not of Aβ40, seed fibril formation. J Am Chem Soc 138:1772–1775
5. DeToma AS, Salamekh S, Ramamoorthy A, Lim MH (2012) Misfolded proteins in Alzheimer's disease and type II diabetes. Chem Soc Rev 41: 608–621
6. Wang ZL (2000) Transmission electron microscopy of shape-controlled nanocrystals and their assemblies. J Phys Chem B 104:1153–1175
7. Kristoffer Bra Nnstro M, Gharibyan AL, Islam T, Iakovleva I, Nilsson L, Choo C, Sandblad L, Pamren A, Olofsson A (2018) Scanning electron microscopy as a tool for evaluating morphology of amyloid structures formed on surface plasmon resonance chips. Data Brief 19: 1166–1170

Chapter 20

Catalytic Efficiency of the Designed Peptides

Kalpana Kumari and Vibin Ramakrishnan

Abstract

This chapter focuses on the estimation of the catalytic efficiency of peptide catalysts. As a working model, a short self-assembled peptide-based catalyst as a novel hydrolase model was constructed to mimic the active site of natural hydrolase. Peptides having imidazolyl group as catalytic center possess high catalytic activity for the hydrolysis of a chromogenic substrate, p-nitrophenyl acetate (p-NPA). In this way, remarkable catalytic efficiency has been achieved by constructing such self-assembled peptides. The de novo designed peptide-based material has great biocompatibility and can potentially be used as a substitute for natural hydrolases.

Key words Enzyme, Catalysis, Active site, Michaelis–Menten constant, Substrate

1 Introduction

Biochemists were greatly inspired by the selectivity and efficiency of natural enzymes [1]. Biologically inspired synthetic catalysts have extensive applications in chemical engineering, environmental sciences, and medicine [2, 3]. Recent advancements in the field of synthetic chemistry led to the development of peptide-based enzyme mimics. Such mimics can potentially be used for regio- and stereoselective synthesis [4]. There are several advantages to use short peptides for catalysis. The catalytic efficiency of such peptides can be modulated by changing the amino acid sequence. Ease of synthesis, stereoselective products, and structural simplicity of short peptides compared to complex native enzymes are other advantages [5–7]. Peptide-based catalysts are, therefore, capable of catalyzing complex reactions, with added advantages that natural enzymes cannot offer [8, 9].

Short peptides offer a direct analogy to enzymes because peptide catalysts have the same or similar constituent amino acids [10, 11] present in their active sites. This chapter outlines the calculation of the kinetic parameters of a peptide catalyst using a hydrolase model by considering p-nitrophenyl acetate as a

Vibin Ramakrishnan (ed.), *Biophysical Characterization of Functional Peptides*, Springer Protocols Handbooks,
https://doi.org/10.1007/978-1-0716-3405-9_20,

chromogenic substrate. It forms a yellow-colored product, p-nitrophenol, after completing the hydrolysis reaction, which has an absorbance (maxima) at 405 nm (Fig. 1) [11].

2 Materials

1. Microplate Reader (Agilent BioTek, Thermo Fisher, BMG LABTECH, Bruker optics).
2. Peptide stock solution.
3. Quartz cuvette.
4. UV-Vis spectrophotometer.
5. pH meter.
6. $ZnCl_2$ stock solution.
7. 96-well plate.
8. Multichannel pipette.
9. p-Nitrophenyl acetate.
10. TRIS pH 8.
11. Milli-Q water.

3 Methods

3.1 Instrument Preparation

1. The instrument should be on 15 min prior to the scan.
2. Set up the kinetic parameters with wavelength 405 nm, 10 s time interval, 400 s of total scan time at 25 °C.

3.2 Sample Preparation

1. Peptide stock and working solution: Dissolve lyophilized peptides into 500 μL of 10 mM of HCl. Determine the concentration of stock by using a UV-Vis spectrophotometer (*See* **Note 5**). Prepare 1.1 mM of peptide stock solution by using TRIS buffer (pH 8). Take 180 μL of this stock solution, 20 μL of isopropanol, and 1.8 mL of 25 mM TRIS buffer (pH 8) and mix it well. So, the final concentration of the peptide working solution is 99 μM (*See* **Note 1** and **Note 2**) [12, 13, 14].
2. TRIS buffer solution: Prepare 25 mM TRIS buffer of pH 8. Weigh 121.14 mg of TRIS and dissolve in 20 mL of MQ. Maintain pH 8 by adding 1 mM HCl. Make up the volume to 40 mL.
3. Zn^{2+} solution: Prepare 1 mM $ZnCl_2$. Weigh 5.4 mg of $ZnCl_2$ and dissolve in 100 μL of 0.1 M HCl. Add 500 μL of TRIS buffer (pH 8). Mix well. Make up the volume to 40 mL.

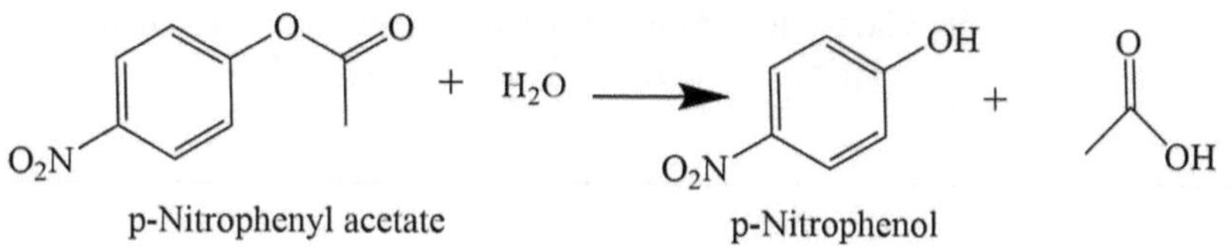

Fig. 1 Hydrolysis of p-nitrophenyl acetate into p-nitrophenol

4. p-Nitrophenyl acetate stock solution: Prepare 0.1 M p-NPA solution. Weigh 90.5 mg of p-NPA and dissolve in 5 mL of acetonitrile.
5. Prepare different concentrations of p-NPA (substrate) to conduct a substrate-dependent study. Consider a series of substrate concentrations ranging from 0.1 mM to 1.2 mM.
6. Add 50 μL of the working solution of the peptide to 150 μL of substrate and buffer.

3.3 Data Collection

1. Kinetics experiments: Collect all kinetic measurements by using a plate reader at wavelength 405 nm at 25 °C in 96-well clear plates. Measure the absorbance of the product, p-nitrophenol formed after hydrolysis of p-nitrophenyl acetate (Fig. 1). Collect data by using MPM 6.3 software.
2. Add substrate at last in the 96-well plate by using a multichannel pipette (*See* **Note 3**).
3. Place the 96-well plate in the holder and start the scan (*See* **Note 6**).

3.4 Data Analysis

1. Export data into an excel sheet and plot the absorbance on the y-axis and time on the x-axis. The addition of peptide (Ac-I-HIHIYI-NH_2) led to the increase in the hydrolysis of the p-NPA substrate due to the formation of more products [12]. This reaction was monitored for 400 s by measuring absorbance at 405 nm.
2. Calculate the initial rate (V_0) by using the following relation:

$$V_0 = (\Delta A / t) \times \epsilon \qquad (1)$$

   where ε is extinction coefficient (= 16,600 M^{-1} cm^{-1} at pH 8) (*See* **Note 7**).
3. Subtract the rate of hydrolysis of the substrate in the blank (without peptide) from the rate of hydrolysis of the substrate in the presence of peptide.
4. Plot the initial rate (V_0) on the y-axis and substrate concentration [S] on the x-axis. Fit this data into Michaelis–Menten equation to calculate catalytic constant k_{cat} and Michaelis constant K_M, where $k_{cat} = v_{max}$/Peptide concentration (*See* **Note 4**).

5. Catalytic efficiency can be calculated by using the following relation, catalytic efficiency = k_{cat}/K_M.

4 Notes

1. Always use freshly prepared samples, including buffer.
2. Prepare all samples by using Milli-Q water and HPLC-grade solvents.
3. Add substrate at last in the reaction mixture and start taking the reading. Because once the substrate is added, it will start the catalysis reaction.
4. K_M is the substrate concentration at which the reaction rate is half of the maximum. Lower K_M represents a stronger affinity toward the substrate and hence more conversion of substrate into product.
5. Calculate peptide concentration by measuring the absorbance A using Beer–Lambert law ($A = \varepsilon Cl$) at a path length $l = 1$ cm. Make a dilution of stock peptide solution in order to calculate its concentration, C [15].

   In order to get the final concentration of stock peptide, multiply the dilution factor (DF = Final volume/original volume of peptide stock solution) and concentration of the diluted peptide solution.
6. Since the path length of the well in the 96-well plate is not fixed, as it depends on the volume of the sample, plate reader, and dimension of the wells. So, always calculate the conversion coefficient, CC (CC = $\text{Absorbance}_{\text{UV-Vis}}/\text{Absorbance}_{\text{Plate reader}}$) before measuring kinetic parameters [16].
7. Use different extinction coefficients of the product at different pH values, for example, 1700, 4300, 9100, 12,700, and 16,600 $M^{-1}\ cm^{-1}$ at pH 6, 6.5, 7, 7.5, and 8, respectively, and 18,700 $M^{-1}\ cm^{-1}$ at pH ≥8.5 [17].

References

1. Korendovych IV, DeGrado WF (2014) Catalytic efficiency of designed catalytic proteins. Curr Opin Struct Biol 27:113–121
2. Bowden AC (2017) Fundamentals of enzyme kinetics. Scientific International, New Delhi
3. Nagar S, Argikar UA, Tweedie DJ (2014) Enzyme kinetics in drug metabolism: fundamentals and applications. In: Enzyme kinetics in drug metabolism. Springer, Cham, pp 1–6
4. Wilson AT (1960) Synthesis of macromolecules under possible primeval earth conditions. Nature 188:1007–1009
5. Rode BM (1999) Peptides and the origin of life1. Peptides 20:773–786
6. Johnsson K, Allemann RK, Widmer H et al (1993) Synthesis, structure and activity of artificial, rationally designed catalytic polypeptides. Nature 365:530–532
7. Zozulia O, Dolan MA, Korendovych IV (2018) Catalytic peptide assemblies. Chem Soc Rev 47:3621–3639
8. Saikia J, Ramakrishnan V (2023) Peptide nanocatalysts. In: De novo peptide design. Elsevier, Amsterdam, pp 173–206

9. Saikia J, Dharmalingam K, Anandalakshmi R et al (2021) Electric field modulated peptide based hydrogel nanocatalysts. Soft Matter 17: 9725–9735
10. Saikia J, Bhat VT, Potnuru LR et al (2022) Minimalist de novo design of an artificial enzyme. ACS Omega 7(23): 19131–19140
11. Rufo CM, Moroz YS, Moroz OV, Korendovych IV et al (2014) Short pept self-assemble to produce catalytic amyloids. Nat Chem 6:303–309
12. Al-Garawi ZS, McIntosh BA, Neill-Hall D et al (2017) The amyloid architecture provides a scaffold for enzyme-like catalysts. Nanoscale 9:10773–10783
13. Verpoorte JA, Mehta S, Edsall JT (1967) Esterase activities of human carbonic anhydrases B and C. J Biol Chem 242:4221–4229
14. Harun R, Danquah MK (2011) Enzymatic hydrolysis of microalgal biomass for bioethanol production. Chem Eng J 168:1079–1084
15. Hollas JM (2004) Modern spectroscopy. John Wiley & Sons
16. Kaspar F, Giessmann RT, Krausch N et al (2019) A UV/Vis spectroscopy-based assay for monitoring of transformations between nucleosides and nucleobases. Methods Protoc 2:60
17. Lengyel Z, Rufo CM, Korendovych IV (2018) Preparation and screening of catalytic amyloid assemblies. In: Peptide self-assembly. Springer, Cham, pp 261–270

MIX
Papier aus verantwortungsvollen Quellen
Paper from responsible sources
FSC® C105338
FSC
www.fsc.org

If you have any concerns about our products,
you can contact us on
ProductSafety@springernature.com

In case Publisher is established outside the EU,
the EU authorized representative is:
Springer Nature Customer Service Center GmbH
Europaplatz 3, 69115 Heidelberg, Germany

Printed by Libri Plureos GmbH
in Hamburg, Germany